POLICY AND PLANNING AS PUBLIC CHOICE

For Christina, Naomi, and Chloe Lewis

and

For Joseph and Geoffrey Williams

Policy and Planning as Public Choice

Mass transit in the United States

DAVID LEWIS, Ph.D.
President, Hickling Lewis Brod Economics, Inc.
and
FRED LAURENCE WILLIAMS, Ph.D.
United States Department of Transportation

LONDON AND NEW YORK

First published 1999 by Ashgate Publishing

Reissued 2018 by Routledge
2 Park Square, Milton Park, Abingdon, Oxon, OX14 4RN
711 Third Avenue, New York, NY 10017, USA

Routledge is an imprint of the Taylor & Francis Group, an informa business

A Library of Congress record exists under LC control number: 99072332

ISBN 13: 978-1-138-33481-6 (hbk)
ISBN 13: 978-0-367-00001-1 (pbk)
ISBN 13: 978-0-429-44508-8 (ebk)

Contents

Contents

List of Figures

List of Tables

Preface

Mass transit is an instutitionalized local public sector function embraced by Federal, State and local taxpayers. A methodical exploration of why the American public so persistently supports transit is long overdue.

To many observers, public sector budgets are a mystery, obscured by arcane technical jargon and unspoken inside information. From this all too accurate perception, some draw the conclusion that public sector budgets are irrational or dishonest. For many, it follows that dependence on public sector budgets makes transit inefficient and ineffective.

The idea has taken hold among many transportation professionals that public transit is a subsidized monopoly which is inherently inefficient. Numerous studies suggest that public transit subsidies have been siphoned off by higher labor costs and inefficient service expansions. Indeed, such transit "efficiency" studies conducted in the later 1970s and early 1980s so impressed transit policy boards and transit managers that nearly 20 years later transit managers still think of little else. A valid concern for efficiency has eclipsed interest in transit's public policy missions.

In the last half century a new body of theory has emerged to examine public sector budgets. The theory of public choice looks beneath the surface rhetoric associated with public policy to find structure and, often, rationality. The application of such public choice concepts as "club" benefits and "spillover" benefits to transit services is a useful strategy for cataloging and measuring transit's diverse benefits.

In Chapter 1 the authors construct a nine-cell "benefit matrix", cross tabulating transit's three public policy functions against three benefit classes that are familiar to public choice theorists. Transit's three functions are low cost mobility, congestion management, and support for pedestrian-oriented neighborhoods and commercial centers. In serving each function, transit produces benefits for fare-paying passengers (market benefits), local taxpayers (club benefits), and general (State and Federal) taxpayers (spillover benefits).

The remaining chapters report on several promising approaches for measuring the monetary value of transit benefits across the nine cells of the benefit matrix. These methods are customarily used by economists to estimate the value of services in the economy. To estimate value from "willingness to pay", from property values in the proximity of transit stations, from substitution savings, and from economic models are practical and intuitive methods that find a ready application in transit.

As the manager of the United States Federal transit program, I have found this particular research useful. The evidence developed here has been used in the Federal budget process to foster renewed discussion of transit's role in the American economy.

To implement recent Federal legislation that has broadened the purpose of Federal New Starts projects, Federal Transit Administration (FTA) planners have examined the benefits research reported here. FTA's continuing efforts include applying the public choice framework into planning requirements and transit planning practices.

I would like to encourage local planners and policymakers to review this material and assess its usefulness in their own practices. We think the public choice framework could be as useful for incremental transit budgets as for new transit proposals. I certainly welcome a lively debate to improve on this valuable work.

The Federal Transit Administration financed and managed the research for this book. Dr. David Lewis and his firm, Hickling Lewis Brod Economics, Inc., performed most of the research under an FTA contract. Dr. Fred Laurence Williams, a long time FTA employee whose own research and conceptual work is presented in these pages, has guided the project from its beginning.

Commercial publication was selected as the most efficient way to broaden the audience and invite dispassionate discussion of the merits. Accordingly, the authors submitted the manuscript to the publisher with no prior review by the Department of Transportation. It is nevertheless true, as mentioned in the acknowledgements, that numerous Department of Transportation employees and independent experts contributed to the research reported in the book.

Gordon J. Linton
Washington, D.C.

Acknowledgements

This work could not have been completed with the cooperation of a number of people over a period of several years.

Foremost, Hon. Gordon J. Linton, Administrator of the Federal Transit Administration of the United States Department of Transportation, provided the firm leadership, financial resources, and staff time to make this study possible. In the course of numerous reviews of the work contained here, Mr. Linton has supplied a consistently persuasive vision of transit's importance in the lives of people and America's cities. His sustained personal encouragement and insistance on the highest research standards has contributed significantly.

Mr. Richard P. Steinmann, Director of the Federal Transit Administration's Office of Policy Development, has worked with the authors on every facet of this research from the beginning. We owe many insights and corrections to Mr. Steinmann's considered judgment. While our most telling critic, Mr. Steinmann has provided a work environment that sustains important new work.

Ms. Janette Sadik-Khan, former Deputy Administrator of the Federal Transit Administration, brought our findings to the attention of the public and the lawmakers who establish transportation policy in the United States. Her ability to distill research, enforce intellectual rigor, and to make sound research useful to policymakers has been critical to our research design. She inspired a devoted and happy ship.

Ms. Charlotte M. Adams, Associate Administrator for Planning of the Federal Transit Administration, was instrumental in launching this research into the valuation of transit benefits. Mr. John W. Spencer, Deputy Associate Administrator for Budget and Policy at the Federal Transit Administration, has steadfastly supported this research and has effectively incorporated its results into the Department of Transportation's performance measures under the Government Performance Results Act. Ms. Grace Crunican, Director of the Oregon Department of Transportation, inspired a strategic perspective that strongly influenced the direction of the research.

The firm of Hickling Lewis Brod Economics, Inc. conducted most of the research for this book. Major Hickling Lewis Brod Economics contributors included Mr. Daniel Brod, Mr. Stephen Lewis-Workman, and Mr. Michael O'Connor. Other important contributors were Morris Davis, Inc., Multisystems, Inc., and Cambridge Systematics, Inc.

The authors would like to thank Kenneth Small, Clifford Winston, Herbert Mohring, Brian Cudahy, Mortimer L. Downey, and Michael Deich for their comments on various portions of the analysis.

This book represents the views of neither the persons cited above nor the United States Department of Transportation or its subdivisions.

1 A Public Choice Analysis of Transit Budgets

Introduction

The professional literature on public transit policies in the United States is incomplete. To contend with chronic financial distress, the preponderance of policy analyses have focused on transportation system efficiency goals, but they have neglected transit's public policy functions. The cost and effectiveness of transit services in reaching transportation objectives (e.g., patronage) imputed to transit systems are well documented. Little, however, has been reported on transit's measurable benefits to passengers and to taxpayers. The ensuing misapprehension of transit's value to households, cities, and the public interest has resulted in a vicious circle of perceived failure and financial neglect in the United States transit industry.[1]

The evidence in this book suggests that the public realizes five dollars in cash savings for each tax dollar invested in transit services. These are the costs of owning, operating and accommodating automobiles that several million Americans avoid with the help of transit services.[2]

As real social processes in which tax revenues are exchanged for transit benefits to taxpayers, hundreds of local governmental budgets have shaped United States transit services since the 1970s. The distinct public policy functions that have emerged from these budgets are discussed throughout this study. How local budgets integrate the benefits of transit is discussed in this chapter. The more pressing question, addressed throughout this book, is how to measure transit's benefits.

Transit budgets are not unusual. Organized vested interests such as construction companies, equipment manufacturers, land speculators, labor unions, and transit managers figure in any local transit policy process. Much about transit can be explained in terms of the pecuniary interests of these familiar groups whose benefits are tied to transit expenditures rather than services per se. To advance their interests over time, however, these interests contend with planners, financial specialists, economists, informed citizens, and other specialists who exert influence on behalf of rationality and efficiency. Professionals wield considerable influence on elected officials who are forced to weigh competing demands on the budget. In addition, public opinion and parochial interests that benefit from transit services indirectly play an important, if episodic, part. In this way, local transit budgets are typical of the budgets served up by polyarchies.[3]

What is a "Public Choice Budget Analysis?"

Decision makers in public service seek to serve the "public good". But how are decisions that serve the public good actually identified and distinguished from publicly bad decisions? Traditional planning theories propose that good public decisions are "rational" in the sense that total benefits to society will exceed total societal costs. The idea is that collective choice can and should mirror "rationality" as it applies to individual choice-making behavior. Individuals do not freely make choices whose costs to them exceed the benefits they perceive to be forthcoming. By the same token, traditionalists argue that social groups in a democratic society should be presented with public choices whose collective benefits exceed the collective costs of achieving them.

In the same vein, traditional neo-classical economics teaches that good public choice requires decisions that yield "Pareto improvements" whereby change leaves some individuals better off without leaving others worse off. Cost-Benefit Analysis and related methods of "rational analysis" are the measurement tools that have been devised to help decision makers make good choices (Pareto improvements) and avoid bad ones.

There are theories of choice however that do not hold to the traditional model outlined above. James Buchanan, founder of the "public choice" school of economics, rejects the fundamental premise that "rational" decision making, as it applies to individuals, can logically and reasonably be transferred to a *collection* of individuals (namely, the public) as a basis for public decision making. Other non-traditionalists, such as political scientists David Braybrook and Charles Lindblom, hold to the same view. Buchanan puts it thus:

> "Rationality or irrationality as an attribute of the social group implies the imputation to that group of an organic existence apart from that of its individual components. If the social group is so considered, questions may be raised relative to the wisdom or "unwisdom" of this organic being. But does not the very attempt to examine such rationality in terms of individual values introduce logical

inconsistency at the outset? Can the rationality of the social organism be evaluated in accordance with any value ordering *other than its own?*"[4]

Buchanan and others of the public choice school argue that it is simply majority decision making in the context of democratic institutions that yields sound social choices. They view majority decision and coalition formation as the key mechanisms through which a social group makes "correct" choices among alternatives, not Cost-Benefit Analysis.

Over the last quarter century, many practitioners of Cost-Benefit Analysis have found little or no evidence that the benefits of transit outweigh the costs. On the other hand, democratic, majority-driven public budgets have supported the provision and growth of transit services in American cities for more than a quarter century. Whereas traditional and neo-classical decision theorists would view the contradiction as an example of bad ("irrational") public decision making, the public choice theorist would assume that weaknesses in the Cost-Benefit Analysis are the cause. Buchanan himself reminds us that decisions reached through the approval of a majority has never been, and should never be, correctly interpreted as anything other than a provisional choice of the social group. As a tentative choice, the majority-determined policy is held to be preferred to inaction, but it is not to be considered irrevocable. In other words, if the result of a majority budget decision to finance transit is ultimately seen by a majority to yield net negative outcomes, the decision will ultimately be reversed. But, if transit budgets are consistently sustained through voter behavior, it is the Cost-Benefit Analysis that must be faulty, not the budget decisions themselves.

In fact, "pure" public choice theorists reject the idea that *any form* of Cost-Benefit Analysis can reasonably inform public decision makers. They take the libertarian position that elected bodies should make decisions and voters judge the consequences of such decisions in the form of subsequent election outcomes. We do not share the view that Cost-Benefit Analysis has no role to play in the budget and decision making process. But we do believe that the long history of democratically determined growth in transit budgets against a trend of Cost-Benefit Analysis results that sought to guide decision makers in the opposite direction indicates that something is wrong with the Cost-Benefit technology, not the public's ability to make choices in its own interest.

Thus the purpose of a "*public choice budget analysis*" is to look to the history of democratically determined budget decisions themselves for evidence of what's missing in the conventional framework for measuring the benefits of transit. Having thus identified the missing elements in conventional Cost-Benefit Analysis, improved measurement modalities can be developed and inculcated into the Cost-Benefit framework. The aim is to sharpen the tools of rational analysis, enabling them to guide decision makers effectively in future. This Chapter provides the public choice budget analysis. Subsequent chapters address the measurement issues.

Transit Services in the Transportation System

Most policy analyes of transit services have abstracted the "transportation system" components of transit services from the public policy functions of transit.[5] Policy analyses have been sharply critical of the transit policy process itself. According to most, transit's "system" role has been worsened by well-meaning, but ill-conceived, public expenditure. The political process is said to interfere with the efficient allocation of transit services. To expand transit's political base, it is said, transit boards have increased peak period commuter services to the suburbs, worsening the inefficiencies of transit's "peaking problem". To appeal to taxpayers, one hears, services are deployed in low-density neighborhoods with hardly any patronage. New systems have been built, it is said, to transform an area of potential high density into a "world class city", an appeal to the public's vanity rather than the facts.

Moreover, analysts of transit expenditures have had only slight success in measuring the benefits of transit services.[6] Like other citizens, analysts naturally search for results in the "sales" or use of the service. Under a planning "postulate" that transportation benefits are to be found "on the network",[7] transit policy analysts looking for "proof in the pudding"—patronage—have found transit policy decisions devoid of rationality.

We believe that the transportation systems framework is, by itself, ill suited to the understanding and evaluation of ongoing public services like transit. Like nearly all public sector activities, transit programs must compromise ostensible "system" goals in serving many masters. In polyarchies, that is, public policies seldom match any particular point of view, not even that of their proponents. Like most urban public policy choices, transit budgets are the offspring of multifaceted urban contention. Like the final score in a sporting event, the outcome of polyarchical decision making exists at a remove from the capabilities of any player or any team of players. Indeed, it is the hallmark of democracy that a typical policy outcome is not a sum of opinions, but a score that settles a contest.

The design and alignment of the Eisenhower interstate highway system in urban areas exemplifies polyarchy outcomes. Originally conceived as a national, limited access high speed road network to link cities together, the interstate highway system was seen by local interests as a way to substitute for locally financed bypass roads. So, instead of staying clear of the inevitable traffic congestion in big cities, the interstate highways were realigned to cut right through the major cities. Highway bills were also "jobs" bills, which encouraged "oversubscription" to the network. Interstate links became "slum clearance" projects and spurred suburban development. In the end, intercity linkage proved to be *primus inter pares* among contending goals.

Stated policy goals such as "national defense highways" are not cynically conceived to "package" nefarious aims. Often the stated goal is the essence of the legislation. But stated policy goals necessarily gloss over the mix of contending goals that are imbedded in most major legislation.

The stated goals of Federal transit programs, the subject of this book, are to improve urban transportation planning and salvage struggling transit service providers in order to preserve cities, combat traffic congestion, and provide low cost mobility to disadvantaged people.[8] While conflicting somewhat, these goals appear compatible.

In the body of transit legislation, however, Congress advanced other goals. The Federal Transit Act of 1964 (as amended in the years since) protects the collective bargaining rights of transit employees, walls off the charter and school bus business, insists on the purchase of United States made transportation equipment, defends the rights of minorities in service and employment, and enforces the financial integrity of Federal grants. These ancillary provisions obviously shore up the coalition of interests that support transit legislation. Clean air legislation adds other transit goals, as does legislation for the rights of people with disabilities, energy conservation and welfare reform.

The commonplace of political compromise should not be mistaken for frivolity. Analysts may object to "ancillary" goals that compromise the legislation's stated goals,[9] but it is important to understand that without the "ancillary" goals from other quarters, transit programs themselves would compromise yet other important public goals. Instead, it is most useful to view Federal transit legislation as a road map of political history, representing the very hard work of navigating public transit goals through a thicket of related legitimate goals, some friendly, some antagonistic.[10] Once subdued legislatively, the public policy goals that may appear to encumber transit programs become a constellation of program support. Thus does transit become a public policy institution that endures.

Looking back at the remarkable stability of local and State transit funding since the 1970s, particularly as the Federal government gradually retreated from operating subsidies after 1982 (discussed below), it is evident that transit budgets have become institutionalized in local political processes. Transit has become as much a function of local governments as education, fire protection, law enforcement, and snow removal. This state of affairs owes much to the diverse "portfolio" of public and private transit clientele, and their multiple goals, among taxpayers.

Cui Bono? A Public Choice Framework for Transit

The transformation of local transit has unsettled many transportation professionals accustomed to an arm's length association with "politics". The budgetary process is accurately seen as something that could not have been designed by an engineer, economist, or even a lawyer. The economist Herbert Simon characterized administrative theory as "peculiarly the theory of intended and bounded rationality—of the behavior of human beings to satisfice because they have not the wits to maximize".[11] "Satisfice" means to set goals that achieve something less than complete satisfaction. "Bounded rationality" and "satisficing" aptly describe most public sector budget decisions in the United States.

The most well-known and exhaustive exposition on American pluralistic decision making was written by David Braybrooke and Charles E. Lindblom in a 1961 book entitled *A Strategy for Decision.*[12] Braybrooke and Lindblom coined the term "disjointed incrementalism" to describe the behavior of representative institutions in the United States and in other constitutional democracies. Disjointed incrementalism is contrasted with the "synoptic model of decision making", which attempts the comprehensive modeling of policy means, ends, and outcomes.[13]

Planning Models as Synoptic Policy Analysis

Synoptic analysis is most applicable to major new investments or discrete decisions having significant costs and local ramifications. Political revolutions that ultimately fail are synoptic in nature, driven by a comprehensive theory of society—a utopian ideal. More practically, the Allied invasion of the Normandy Beaches in 1944 and the United States

Space Program exemplify major achievements in synoptic analysis on an historical stage. Normandy invasions and Moon landings required unusual budget decisions. Conceived and designed to be within the capability of the relevant agents, the principal goals were not allowed to be compromised by other goals.

In the transportation sector, the "alternatives analysis" the Federal government requires for major new transit investments is an excellent example of synoptic analysis. Major transit projects are analyzed comprehensively. The project itself is elaborately modeled with data on travel times, population demographics, local economics, and other factors relevant to the project's transportation benefits and goals. Alternatives analysis also includes extensive modeling of environmental effects on neighborhoods, habitats, and public safety. The analysis includes costs and benefits as well as strategies to mitigate detrimental impacts.[14] Even long term financial support is incorporated in the modeling process. Alternatives analysis establishes the technical parameters for decisions on major transit investments in the United States.

The alternatives analyses used by transit planners serve as the empirical foundation of the projects in the decision making process. However, transit models often run aground in the unseen shoals of the budgetary process. Few public policy decisions are decided on technical parameters alone. Nor should they be. According to Braybrooke and Lindblom, the synoptic model is flawed as a model of most decision making because *adjustments* to the idiosyncrasies of specific public policy environments simply do not yield to systematic analysis. In making policy and budgets, decisionmakers put these adjustments at the very center of their calculations as they reach agreements across the spectrum of diverse interests and points of view that constitute the policy process.

Braybrooke and Lindblom's "disjointed incrementalism" pursues meliorative goals, fixing problems that emerge in the political agenda by comparing the status quo in very limited terms with changes at the margin—and whenever possible postponing further fixes. The "problems" usually correspond to competing legitimate interests and conflicting objectives held by the community as a whole. Braybrooke and Lindblom briefly describe "disjointed incrementalism" as follows:

> "Several features of incremental or margin-dependent choice need to be distinguished. . . . First, only those policies are considered whose known or expected consequent social states differ from each other incrementally. But one can imagine that a set of policies meeting this condition might be expected to bring about some social state differing

drastically and non-incrementally from the status quo. Hence we add a second feature: that only those policies are considered whose known or expected consequences differ incrementally from the status quo. The third feature of incremental choice, however closely it seems to follow from the first two, is logically independent of them: That examination of policies proceeds through comparative analysis of no more than the marginal or incremental differences in the consequent social status rather than through an attempt at more comprehensive analysis of the social states. To this list we add a final feature, again logically independent but implicit in our exposition of incremental choice: choice among policies is made by ranking in order of preference the increments by which social states differ".[15]

Thus, disjointed incrementalism strongly suggests the notion that public decisions are best made on the basis of cost and benefit information at the margin, under the cross pressures of competing budget priorities, in the same way that marginal cost pricing drives Pareto optimal economic decisions for the firm or household.

The distinction between disjointed incrementalism and the ideal of synoptic modeling is very important for understanding the benefits that transit budgets create. Most existing policy analyses of public transit are premised, perhaps unwittingly, on the synoptic ideal. They judge transit performance by standards embedded in "technical" transportation analysis. Transit patronage and net cost per passenger mile lead most analyses. Some analysts argue that any benefits should be traceable to patronage. The more cost effective transit is, the argument goes, the more it can contribute to solving "the" urban transportation problem. When viewed from the perspective of the taxpayer or the elected official, synoptic analyses of transportation policies and budgets are "reductionist" in nature, forcing complex goals into easily measured outcomes such as patronage and deficits. These outcomes are important, but they obscure the benefits of transit to the taxpayer. Technical outcomes gloss over the contested nature of public choice, a contest as much within the breast of the individual taxpayer as between competing interest groups.

Patronage as a Budget Factor

A digression on the budgetary significance of patronage is instructive. Bluntly, by and large, patronage is sought to earn and to supplement tax subsidies,[16] and is one indicator that points to genuine policy goals. Three familiar transit phenomena demonstrate that other goals routinely trump patronage. *First*, most large cities have some transit routes that are able to cover most or all of their costs from passenger fares. These routes have high off-peak patronage and bi-directional travel demand. Most are crosstown and local services. They tend to serve high density areas or, in smaller urban areas, large college campuses. Equally common in the same systems are routes that fail by this standard. Why are resources not shifted to the more economically "viable" routes? Politics? Perhaps a "patronage maximizing" strategy to shift services to the highest performing routes is simply "bounded"[17] by the obvious consideration that all taxpayers are entitled to a modicum of the services they pay for. Minimal coverage is not a lamentable circumstance (Politics!), but the product of deliberative democracy. Taxpayers don't pay for patronage, they pay for service.

Secondly, national survey data in 1995 indicated that 27 percent of transit passengers cannot find a seat when they board a transit vehicle.[18] To bypass congested highways, two million commuters pay fares daily for standing room on buses and trains. Transit crowding represents an obvious opportunity to increase market share. In these crowded corridors, where transit is literally drawing more than its share of passengers, more transit service could increase transit patronage and would probably reduce travel times for motorists by dampening highway travel. But these systems do not add capacity. Why? In highly peaked transit operations, revenues at the margin are exceeded by incremental costs, which raises the need for public subsidies. Increasing peak capacity, that is, would increase the ratio of incremental transit costs over revenues in excess of a limit established by the policy process. Thus, crowded commuter trains are the deliberate outcome of a "satisficing" transit strategy, balancing congestion relief against incremental subsidies.[19]

So, it appears that beyond a certain cost limit, transit boards are willing to tolerate discomfort for existing passengers and do not always seek more passengers or market share. Every day, transit boards make conscious decisions to turn passengers away in pursuit of policy goals in which patronage is secondary.

The *third* and most revealing fact is that hundreds of local governments have institutionalized transit despite its persistent failure to increase patronage since about 1980. More than 10,000 annual transit budgets have been adopted by local and State governments across the United States over a 25 year period.[20] Many of these budgets required pooling of line items among local governments that are jealous of their budgetary power. Something other than patronage persistantly and pervasively comes first. Finally, budgeted subsidies per passenger are greatest in smaller urban areas[21] where transit's role is ostensibly least important in terms of patronage and market share.

Transit and Planning Ideals

The criticism toward transit is uniformly grounded in transportation planning models rich in transportation criteria, but poor in hard to measure objectives and benefits. Policies are characterized as being "simplistic" and "shortsighted", not taking into account the complexity of the transit industry or of the transportation problems—like traffic congestion—that transit is purported to address.

In support of the contention that policy makers have been naive about transit, analysts have cited publicly stated promises that transit would not require operating deficits.[22] One well-known research project consisted of contrasting the forecast patronage for a number of transit projects and the disappointing actual patronage years later—proof that project advocates skewed their forecasts the quest for Federal grants. The following passage is illustrative:

> "By tolerating pervasive errors of the consistent direction and extreme magnitude documented here [exaggerating patronage and minimizing costs], the transit planning process has been reduced to a forum in which local officials used exaggerated forecasts to compete against their counterparts from other cities to obtain Federal financing of projects they have already committed themselves to support, but realize cannot prevail in an unbiased comparison to plausible alternatives."[23]

Here the author, with no evidence one way or the other on rail transit benefits apart from those reflected in "new riders", implies that the goals that local officials "have already committed themselves to support" could not stand public scrutiny. Earlier in the same paper he calls them "other—often unspoken—reasons".[24]

Meyer and Gomez-Ibanez in 1981 contended that governments supported transit principally to reduce traffic congestion. Because the complexity was underestimated, the authors said, transit only enabled commuters to live further away from their jobs in the city and had no effect on traffic congestion. As if to summarize their synoptic perspective, Meyer and Gomez-Ibanez suggested that "for the automobile as for the broader issues, coordination, comprehensiveness, and consistency are apt descriptions of what is required. Proliferation of policy goals in recent years has only heightened these needs".[25] In their words: no more "highly simplified diagnoses, without any recognition of the interactions among various urban transportation policies and other public policy goals". No more public policy shifting "from one simple panacea to another".[26] Alan Altshuler in 1979 attributed the "failure" of transit and other transportation strategies to "solutions in search of a problem".[27]

With more research, these writers might have discovered that Federal guidelines and synoptic folkways enjoined the introduction of hard to measure but very large benefits only indirectly related to "net new riders". Confined by the Federal planners' narrow benefit focus that was in force during the period before Congress broadened the class of transit benefits in 1991, local authorities had little choice but to manipulate the few "eligible" goals available in the process.[28]

A 1988 survey of urban planners in Atlanta suggested that the rapid transit system included the goals of combating traffic congestion. But MARTA was also built to help restore the vitality of the central city by facilitating mobility across the region. Rapid transit enhanced Atlanta's self image as an emerging "world class city" (that would later host the Olympic games). Hard-to-measure land use benefits were anticipated as were social benefits associated with regional economic integration.[29]

Specialists in every endeavor find policy processes deficient. Policymaking is untidy. Public support requires the simplification of complex data. Exaggerated promises and rosy scenarios in support of budgets are often necessary to offset equally dishonest tactics on the other side of every issue. It is for this reason that students of public policy are instructed, not to ignore symbolic politics, but always to look for the substance beneath the rhetoric. Slogans that seem empty to the transportation experts can be clues to genuine policy goals implicated in transportation budgets.

A good example of the clash between planner and policymaker is continuing controversy over the 1964 legislation that established a full-fledged Federal transit agency. The original Federal Transit Act was preceded by a two-year process in which most facets of the transit industry were thoroughly discussed in the United States Department of Commerce, in Congress, and in the media.[30] Yet one historian stated that the United States Congress "backed into a deeper commitment to mass transit without reaching agreement on the objectives to be achieved and without anticipating the eventual cost of federal involvement".[31] The same commentator also reported that earlier Federal transit legislation in 1961 was "an opening wedge", after which "transit lobbyists" marshaled "support for an urban renewal-style capital grant program" that had always been the urban-rail alliance's long term goal.[32]

The same commentator characterized the generic local transit budget process:

> "The drama of shoring and patching has been played out in city after city in the era of public ownership [of transit]; It can be characterized as follows. With stable fares and rising labor costs, passenger revenues are unable to keep pace with the growth of the property's wage bill. A fare increase or service reduction must be considered. The discussion of increased fares mobilizes public reaction, generating civic and political pressure for a 'better alternative'. One such alternative is increased subsidy. A proposal to increase subsidy is received skeptically, but it gradually emerges as the preferred option through artful packaging. The packaging that proves acceptable includes a relatively small increase in subsidy, a mild fare increase, improvements in peak-hour commuter service, and reductions in poorly patronized off-peak service. The combination of service improvements and service reductions is presented as a rationalization program, but its net result is an increase in the property's operating and capital cost obligations because of the high cost of additional peak-hour service. Passenger revenues increase proportionally less than do newly occasioned costs, and thus the financial cushion afforded by additional subsidy and higher fares is exhausted in short order. The deftly packaged plan that had been presented as a long-term solution to the property's financial woes proves durable for only a short time—perhaps two or three years. Management recognizes that a precipitous new request for additional subsidy would be rebuffed; further economy measures must be taken first. Thus, nighttime and weekend services are reduced yet again, producing modest savings. More important than their magnitude these economies mobilize no influential constituencies and, more important still, symbolize and communicate management's resolve to control and reduce costs. Management has shored and patched; the property has demonstrated its resolve to cut and prune; it is positioned for the cycle to begin anew".

An account of transit policy making perhaps. But the implied conclusions are not true. Budget processes in local governments commonly focus on marginal changes to the status quo. A fragmented perspective does limit the comprehension of the problems facing transit. Transit advocates do contrive to shape their policies to win public support. And legislative sponsors do indeed have ulterior aims. The very term "incrementalism" implies that legislation is one step on the path toward an ulterior aim.

But it is not true that public support for transit is based on little more than "deft packaging" of transit budgets. Policy makers are keenly aware of the deficit implications of their policies. Rather than the dithering implied above, the discipline of pluralistic politics weighs the economic facets of decisions very carefully and does so in a sustained and highly sophisticated fashion. Moreover, "opening wedges" and "ulterior designs" are the daily stuff of legislative debates and public hearings and were certainly raised when governments initiated public transit programs. Aware of the staying power of budget line items, legislatures seldom "back into" legislation.

The contrast between a synoptic and incremental view has been revealed in courts of law where imputed "intentions" have encountered the pluralistic decision making process.[33] To resolve legal conflicts, the adversary process is often able to pry open the more substantive forces at work in legislation and executive decision making. Civil rights litigation is particularly fruitful for uncovering the policies behind the "politics".

Under United States Civil Rights law, transit services or benefits may not discriminate against people because of their race, color, or national origin. A case in New York City involved a fare increase for suburban commuter rail passengers that was lower than that adopted for downtown subway and bus riders. The plaintiffs' argument was synoptic in nature. They argued that this was a case of discrimination against the central city, with its higher proportion of residents who are minorities. The transit authority replied that conflicting policy issues were involved. An appeals court ruled that the transit authority was entitled to charge a different fare because the transit authority demonstrated its case that commuter rail subsidies benefitted the city and its inhabitants, e.g., discouraging suburban residents from driving into the city, arguably offset the fare difference.[34]

In a similar proceeding, a Los Angeles complaint alleged that a new rapid rail system was proceeding at the expense of transit services for low income minorities who depend on bus services. In this case it was the transit authority that represented a synoptic view, seeing little relation between the internal planning of its rail system and the larger problems of society. The transit authority was directed by the court to look again at its larger role in the community and to take specific measures (restore special fares, reduce bus crowding, reduce transfers) to meet low cost mobility needs before completing construction of the rail system.[35]

Another example was a Philadelphia case in which the regional transit system's fare subsidy allocation had the effect of cross subsidizing middle class suburban residents from the fares of inner city low income minorities. Conceding the cross subsidiy effects, the transit authority defended its fare practices as necessary to stabilize regional fares, services and budgets. In addition, other policy options that might mitigate the inequity depended on jurisdictions beyond the transit authority's control. This case illustrates the complex nature of transit's objectives while also drawing attention to other masters outside the authority's control.[36]

In summary, public transit services in the United States are not businesses or corporations. Most that once were businesses went bankrupt many years ago. They are not economic entities operating in a market. Motorized travel in the United States is largely immune to market forces. Public transit services are instruments of public policies that are influenced by transit customers, constituencies and the voters at large. The analysis of transit as a business, or a service operating in a market, exercises a certain appeal because it allows the use of the familiar measures of performance: costs, revenues, sales, etc. Predicably, such analyses find transit services wanting as an economic proposition. These tools of measurement, however, are inappropriate.

Transit Budgets and Budgetary Incrementalism

The "politics" of transit budgets yield to systematic investigation. Students of political science and economics seek out the underlying uniformities that reveal themselves over time in public sector budgets. The democratic form of government prevails with slight variations across all local governments in the United States The object categories upon which these government spend tax dollars are also fairly uniform: school teacher salaries, police salaries, library books, sidewalk repair, street lights, etc. These uniformities are fairly obvious, and Americans take them for granted. There are less obvious, but important, uniformities in the functioning of local governments. To use a familiar example, since mid-century, the scope of public education in the United States has broadened to include health education, student nutrition, active enlistment of parents in the child's education, remedial programs, English as a second language, and numerous other areas of child development. These accretions of functions are institutionalized responses to circumstances that are replicated over time in classrooms and communities across the country. Any major adaptation by a public institution, by law, must be applied generally in the jurisdiction. The functions converge across jurisdictions because the constituencies they address exist across jurisdictions. The institutionalization of governmental functions occurs in the budgetary process.

Legislation and budgets that appear incoherent to one point of view are more appreciated by looking deeper into the legislative process. As suggested above, the analyst needs to look behind rhetoric and legislative language, since these are often simply tools in the political tool-kit. Harold Lasswell, a seminal thinker in the field of Political Science, urged his students to look beyond symbols to the organized interest groups, to latent publics, and to the general public and to ask: "Who gets what, when, where, and how?"

Transit policy analysis should look at "who gets what" (*cui bono*) in two distinct senses. Studies of "special interests" that receive direct financial benefits from public budgets are a staple of serious policy research, investigative journalism and sensational novels. As mentioned earlier, groups realizing capital gains, employment, and commissions for public works are commonplace in transportation decision making. Pressure groups and lobbyists are a central fixture of pluralism, a much lamented necessity in a modern society. But if we judged policies on the basis of how many jobs they appeared to "create"[37] there would be no end to public works, a curiosity most analysts would deplore.

However, Lasswell's "who gets what" criterion also means looking at the benefits of *services* to households at-large, to specific population groups, and to the general public. Systematic examination of these "policy" benefits yields information on the implicit "revealed preferences" of the society as a whole for goods and services that are procured through public sector budgets. In the case of transit benefits, of course, they are jointly procured by passenger and government budgets.

The study of numerous government budgets over the long term suggests underlying public demand for budget consequences. Furthermore, the large span of time and variety of circumstances tends to cancel out the capriciousness of "pecuniary" interests that tug at budgets in every locality. The idea that transit serves "policy functions" is equivalent to predicting that the discrete benefits of transit recur in similar material circumstances.

The buy-out of private transit companies by local governments during the 1960s and 1970s was entered into only reluctantly after years of transit decline and the exhaustion of other ways to preserve low fare transit services.[38] The problems were rather straightforward: fare increases needed to pay transit expenses chased off patronage and crippled the ability of transit to offer basic mobility, to combat increasing traffic congestion, and to maintain high service standards in city neighborhoods and commercial centers. At 67 percent, the Federal share of transit buy-out costs and fleet modernization was a generous spur to local governments. In fact, Federal financial assistance transformed a major new budget crisis for many local governments into a more incremental budget adjustment.

From the beginning, the costs of operating transit services far exceeded capital costs. It was well known that spreading auto ownership and sprawling residential and commercial development deeply undermined transit's ability to cover its operating costs from fares. So it was well understood by local legislators that large transit subsidies would be necessary indefinitely. As early as 1973, the State of California adopted legislation that enabled local communities to dedicate ¼ percent sales tax revenues for transit subsidies. State assistance for operating subsidies was common throughout the United States by the early 1980s.

Evidence for Transit Incrementalism

After the initial years of growth in the 1970s, the national aggregate transit budget stabilized after 1980, tracking closely the general inflation rate. The trend in State and local funding from 1979 to 1995, using constant 1992 dollars, is illustrated in Figure 1.1. The increase was 214 percent.

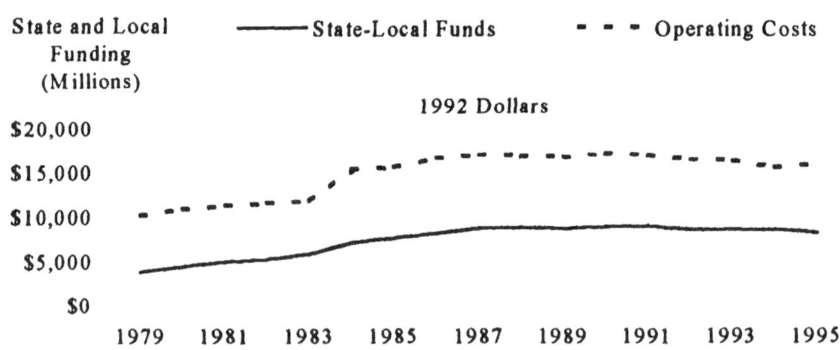

Figure 1.1 Trend in United States' State and Local Transit Funding, 1979 – 1995

Source: Federal Transit Administration, *National Transit Database.*

The incremental character of local transit budget processes is reflected in Figure 1.2, which compares the combined State and local transit contributions individually for 364 transit systems in 1992, 1993, 1995. The solid diagonal line arrays the State and local amounts budgeted for 364 transit budgets in 1992. The average annual change in combined State and

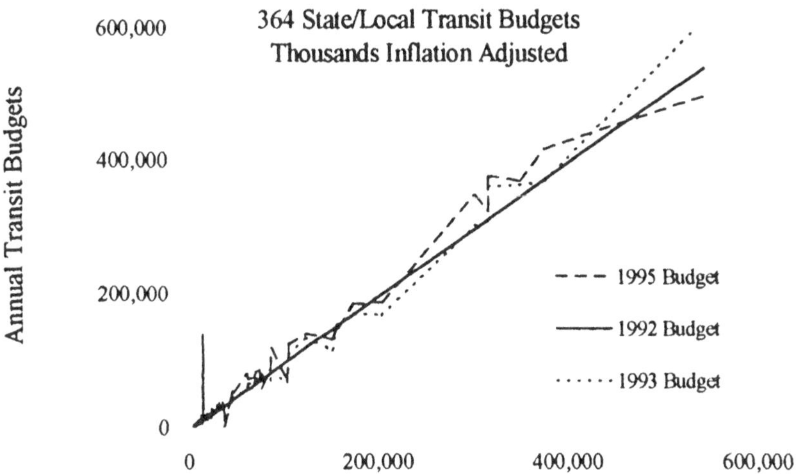

Figure 1.2 Incrementalism in Transit Budgets, 1992 – 1995

Source: Federal Transit Administration, *National Transit Database*.

local transit funding from 1993 to 1995 (2 years) was less than 10 percent for 60 percent of the 364 transit systems reporting. But continuity is only part of the power of incrementalism. Incrementalism is even more powerful a budgetary rule than Figure 1.2 suggests.

The Institutionalization of Transit Budgets

The United States is a growing society in several respects. Local budgets tend to grow with population and tax revenues as well as with inflation. Figure 1.3 shows the per capita "transit tax burden" from 1980 to 1992. The "transit tax burden" is defined as transit subsidies as a share of the per capita general budget. Figure 1.3 shows a one-third decline from 1980 to 1992 in the overall United States transit tax burden. This is the net effect of a two-thirds decline in the Federal transit burden and scant change in the State-local transit tax burden. Figure 1.3 demonstrates how, through incremental budgeting, local taxpayers supplanted Federal budgetary support for transit, without increasing transit's per capita share of local and state budgets. This illustrates how governments institutionalize a new activity. They nearly doubled State and local transit funding while transit budget increments on the whole did not exceed annual increments in the general budget.

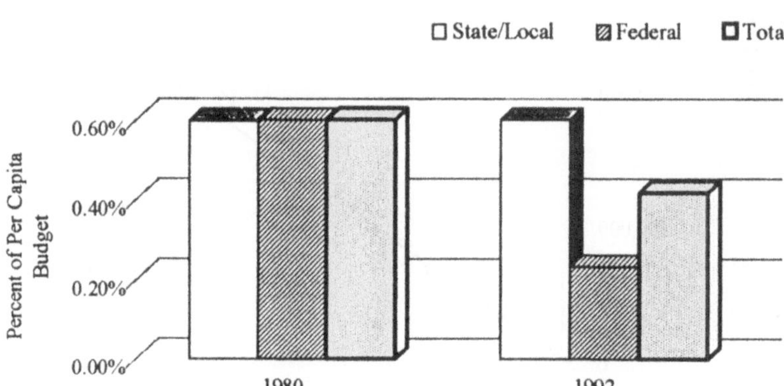

Figure 1.3 Comparative Transit Tax Burden, 1980 – 1992

Source: Federal Transit Administration, *National Transit Database* and Department of Commerce, *Statistical Abstract of the United States* (Washington, D.C.: 1994).

The public establishment of transit services in the United States was part of a larger evolution in transportation. In the decade after the consolidation of Federal transportation programs in the United States Department of Transportation in 1967 the States created their own transportation departments. Certified Metropolitan Planning Organizations (MPOs), consisting of local government representatives, grew up alongside Federal transit grants. The inter-local character of transit routes has led to increasing regional transportation cooperation. Clean air milestones in many cities have added urgency to long term regional transportation planning. More recently, land use development has come under the watchful eye of regional transportation planners. Consultation with local communities has become the watchword for transportation planning under the Intermodal Surface Transportation Efficiency Act of 1991 (ISTEA). These institutional changes in transportation have reinforced the integration of transit into local budgets.

Incrementalism, Rationality and Efficiency

Transit budgets may appear complex on the surface, but over time they reveal patterns of efficiency and equilibrium. When analytical models are introduced in the policy or budget process, their acceptance relies heavily on the voluntary suspension of disbelief in the simplifying assumptions, generalized correlations, and historical abstractions that are taken for granted by technical professionals. The deliberation that occurs in local budgetary processes tends to bring preferences into play with an immediacy rarely captured in formal analysis. The parliamentary dialogue itself shifts weights among preferences. Process substitutes for uncertain or incomplete data. The insertion of community values, skeptical dialogue, and the deliberative processes take a toll on the decisionmakers' indulgence for technical uncertainties. The fanciful account of a "patchwork" policy process cited earlier reveals the frustration many social scientists experience "because they look on those practices in the false light shed by unsuitable ideals of evaluation method".[39]

Implicit in even the most rigorous policy analysis is traditional welfare economics which takes the public's values as "data" that cannot be altered by the decisionmaikng process itself. Professor James Buchanan, founder of the "public choice" school of decision theory, points out the flaw in this reasoning: "The definition of democracy as 'governments by discussion' implies that individual values can and do change in the process of decision making".[40] Noted economist Martya Sen is even more direct: "The practical reach of social choice theory, in its traditional form, is considerably reduced by its tendency to ignore value formation through social interactions. [Many] of the more exacting problems of the contemporary world—varying from famine prevention to environment preservation – actually call for value formation through public discussion".[41] Charles E. Lindblom's name for this process of value change via decision making is the "formation of volition".[42]

Chapter 6 of this book elaborates on the problems of confrontation between cost benefit analysis as currently practiced and the formation of volition in the policy process. In is necessary now, however, to anticipate that later discussion in order to spell out more fully the conflict between prevailing policy analysis and the approach used in this book.

Extremely costly controversies arise from the uncritical extrapolation of individual utility maximization to the procedures of policy analysis and public sector decision making. In a bizarre turning of the tables, pluralistic decision making is perceived as being arbitrary by experts who have internalized the fundamentally mistaken belief that cost benefit analysis is itself the best means to achieve democratic outcomes! Others would characterize this conflict differently, but it is a practical and even heated controversy than demands a digression.

Long ago Kenneth Arrow proved,[43] though his "impossibility theorem", that in trying to obtain an integrated social preference from diverse individual preferences, it is not in general possible to satisfy even mild-looking conditions that would meet the most elementary standards of reasonableness for public choice in a democratic society. Arrow wanted to prove that a social welfare function could satisfy, simultaneously, the following four conditions:

Create a rank ordering of public priorities for every possible combination of individual preferences. ("universal domain")

Allow the ranking of any two social states to depend on peoples' preferences, only over that pair of alternatives, with no dependence on how other, unrelated alternatives, are ranked. ("independence")

Permit no individual or group of individuals to prevail over the social ordering regardless of what others prefer. ("nondictatorship")

All the group of all individuals, taken together, to prevail over the social ordering. (Pareto optimality)

In the end, Arrow proved the opposite. The "impossibility theorem" is a logical proof that it is not feasible to have a social welfare function satisfying, simultaneously, independence, the Pareto principle, and non-dictatorship.

For our purposes, Arrow's proof shows that it is impossible to make a leap from individual preferences to agreed upon societal preferences while still preserving some basic axioms of rationality. For example, sovereign consumers in the market place exhibit "transitivity", when they prefer x to y and y to x, they also prefer x to w. In other words, if a person prefers a faster over a safer trip, but values safety more than highly than clean air, the same person will also prefer speed over clean air. Cost benefit analysis for public policy should not make the same claim, because the logic of transitive individual preference breaks down in groups. As a member of a democratic group, the same person just mentioned willingly supports environmental policies that produce lower speed limits.

Does this mean that group choices are inherently antidemocratic, or elitist, or irrational? In a seminal essay on social choice, Buchanan finds the fault line not in the foundations of democracy, but in the "first principles" from which the impossibility theorem springs. The dilemma lies there, in the substructure of welfare economics, not in the ability of society to make rational choices through democratic (non-dictatorial) means.

> "Rationality or irrationality as an attribute of the social group implies the imputation onto that group of an organized existence apart from that of its individual components. If the social group is so considered, questions may by raised relative to the wisdom or unwisdom of this organic being. But does not the very attempt to examine such rationality in terms of individual values introduce logical inconsistency at the outset? Can the rationality of the social organism be evaluated in accordance with any value ordering other than its own?"[44]

In simple terms, different concepts of "rationality" apply to a whole society as distinct from a single individual. In seeking to impose on cost benefit analysis the logic of traditional welfare economics, and with it one arbitrary notion of "rational" behavior, governments exercise decision making power in a way that necessarily has the appearance of being arbitrary.

Finally, Buchanan goes on to show that the breakdown of transitivity at the collective level is not an obstacle to rational choice but merely an artifact of the assumption that the logic of individual choice is a "good thing" for social groups as well. Judged in this way, even voting can lead to "irrational" decisions. But Buchanan demonstrates that such "irrationality" is actually a *desired* attribute of social choice. Philosophically and historically, majority decision evolved as a practical means to make a group decision when unanimity among group members could not be achieved. Majority rule is for breaking stalemates when a course of action for the group is needed. That is, a majority decision is far from ideal, but is rather a provisional or experimental choice for the group and should never be interpreted as anything more. Thus, the importance of free dissent! Fundamentally, a majority choice is preferred to inaction and nothing more.

Accordingly, logical consistency as a test for acceptability of a policy, while highly valuable, would forclose majority rule even in reaching provisional choises as they arise from day to day in legislatures, courts, and executive agencies. Indeed, policy analyses that, over time, find themselves pitting analytical outcomes against pluralistic policy outcomes are highly desirable as technical tools and as sources of policy dissent and eventual correction. However, the substitution of real policy outcomes with planning models would be highly problematic, and not only from the democratic perspective.

To make the point historically, it is quite literally true that political revolutions quickly take a dictatorial turn precisely because the "new theory" inevitably is a poor fit for the problems of the day. In fact, however widely supported and however right for the times, revolution itself becomes the predominant *problem* for the times. Examples of this social "Murphy's Law" in transportation and other infrastructure litter the landscapes of cities and nations worldwide. It has taken decades to undo abysmal results of post war large-scale low income housing policies in major United States cities. Groaning Third World debt is a monument to grand economic models put into practice without the sobering effects of pluralistic policy processes. The "appropriate technology" movement in developing economies was a response to bring pluralistic processes to bear on developmental planning in underdeveloped economies.

The United States has been blessed by the capacity of its people to coalesce and to force pluralistic processes (often by court order) onto urban transportation planning processes. Indeed, under Gordon Linton's leadership, the U.S. Federal Transit Administration has joined with local interests to bring pluralism deeply into urban transit planning.

The Transit Record

The economic history of transit services in the last quarter century has been the extension of discrete transit benefits to taxpayers while weathering incessant criticism from the synoptic policy analyses of writers cited earlier. The "shoring and patching" of policy that so distracts the analysts was the way budget processes brought information to bear on decisions. Facts were determined to be pertinent, and to warrant assigned weights, on the basis of comparisons at the margin. These were comparisons of costs, of benefits, of revenues, of public demands, and of long term goals, always in increments at the margin.

By contrast, synoptic or modeling approaches that penetrated transit's internal economic contradictions, while interesting, were of limited help for changes at the margin. Synoptic modeling too often was uncertain on interesting long term and systemwide issues and mute on critical issues facing decisionmakers in the challenge at hand. In Braybrooke and Lindblom's words:

> "Is all this merely to repeat the obvious—that if analysts cannot agree on decisions, some kind of agreement will emerge in the political process?

> No, the point goes beyond that. We are saying that where analysis and policy-making are serial, remedial, and fragmented, political processes can achieve consideration of a wider variety of values than can possibly be grasped and attended to by any one analyst or policy-maker. It is this accomplishment at the political level that makes agreement among analysts less necessary".

Incrementalism can be analytically superior to modeling. The annual recurrence of the transit budget offers repeated opportunities for cost implications to come into view and for overly optimistic predictions to become discredited. Each successive transit budget is introduced into a political stream that differs from its predecessor in the mix of political actors, the availability of revenues, the state of public opinion, and the accumulated expectations and experiences with the transit system's performance. This evolving mix brings forth new facts and shifts the weight among already known facts. Eventually, the process "learns" the economics of transit, demographic influences, and the other important dynamics.

If this is true and public policy is Darwinian rather than Biblical, what are transit's survival skills in the legislative process? If not patronage or efficiency, what is the proper performance measure of transit? What have transit services done in 10,000 local budget years that cannot be done by cheaper, faster, more convenient, portable personal cars? What benefits do transit services convey to taxpayers? Are these transit benefits worth $20 billion per year?

Modern Transit Services in the United States

In the spring, 1994, the Chairman of the House Appropriations Subcommittee for Transportation posed a familiar question to the U. S. Department of Transportation leadership: "As a threshold we want to see if you can help us rationalize why if transit use is declining, why we should be spending more in transit. Aren't we ending up paying more and getting less?" This perennial congressional inquiry commands a thoughtful answer. The short answer to Congress follows.

Patronage is important in controlling inevitable deficits in the quasi markets where transit is deployed. But patronage is only incidental to the sizable benefits that local passengers and taxpayers call upon transit to generate. As this book will show, the public and private benefits of a

$17 billion transit industry in 1990 probably exceeded $50 billion. Patronage lowered the taxpayer cost of those benefits to $10 billion.

The longer answer is provided in the Chapters that follow. These Chapters contend with the historical difficulties planners and economists have faced in measuring transit's benefits to American society. Like other forms of economic infrastructure such as education and law enforcement, transportation facilities cast a wide spectrum of outcomes over a long period of time. Only recently, for example, have scientists attempted to catalogue the social costs of automobile travel, and the effort is an extremely difficult and expensive enterprise (discussed in Chapter 2). Some very large long term outcomes, such as land use effects, are the most difficult to trace.

The authors believe that most studies to date have implicitly devalued transit's benefits without having tried to attach monetary values to these benefits. For example, a 1985 General Accounting Office (GAO) study of Federal transit support had the following to say about mobility for the elderly, people with disabilities, and low-income persons:

> "The Congress has expressed specific concern for the mobility needs of elderly, handicapped, and low-income persons who are unable to afford to drive an automobile. Research we examined and local Metropolitan Planning Organization (MPO) officials in the five cities we reviewed generally indicate that the transportation disadvantaged, along with the general public, have benefited from various transit improvements, including increased service levels, improved equipment, and stabilized fares. Also, through special half-fare programs and services such as wheelchair lift-equipped buses and special paratransit services, the federal transit program has helped address special mobility problems of the elderly and handicapped. Research, however, has raised questions concerning whether such approaches have effectively and efficiently addressed the needs of those requiring mobility assistance. For example, some research suggests that targeting subsidies for low-income riders, rather than subsidizing all riders, would more equitably benefit those who need transit assistance."

Although the GAO cited the number of passengers and taxpayers who may have benefited from transit, and expressed concern over the efficiency and effectiveness of transit in providing benefits, the GAO did not address the monetary value of transit benefits to transit passengers and to society at-large.

Certainly, there may be more efficient ways of targeting mobility to the poor, for example. Furthermore, a cost analysis is technically correct without the valuation of benefits. But useful policy analysis cannot be satisfactory without an effort to attach a monetary value to the benefits of existing and desired transit services. Neglecting the benefit side simply leaves the taxpayer knowing "the cost of everything and the value of nothing".

Transit's Role in American Society

As the United States contemplates its transportation and telecommunications networks over the next decades, the public debate over public transit's role in America's urban areas is stalled. Public support for transit services surged in the 1970s led by massive increases in Federal and State financial aid. By allowing the public purchase of transit companies, the influx of public funds rescued the transit industry from secular decline in services and patronage. Patronage rebounded and then stabilized at the 1980 level, where it has remained ever since, about 7.5 to 8 billion trips per year. Transit's "stability" in the face of growing transportation demand reflects the stand-off in transit funding since the early 1980s in perennial municipal budget processes throughout the country, in most State capitals, and in Washington, D.C.

Transit policy is stalled because the public is without reference points. We have been inattentive to the basis of local public support for transit services. Therefore, unlike many studies of transit, the aim of this book is not to argue why taxpayers should support transit, but to discover why taxpayers do support transit to the tune of $17 billion per year nationwide (in 1993). For through local budgets, the public's support for transit has shaped transit and its performance in the United States If we can figure out where we have been in local budgetary processes, we can guess where the same processes might lead in the future.

Many view transit as a flop. Relatively few Americans use transit services. FTA estimates approximately 35 million Americans (14 percent) of the United States population use transit services consistently or intermittently in a given year. Only one in twenty workers ride transit to work. Only one in fifty trips are made on transit.

Yet, the record of public support is impressive. Taxpayers consistently support heavy subsidization for transit and do so in a great variety of circumstances. As shown in Figure 1.4, after 1983 Federal assistance to transit started a steady decline in real dollar terms. In the ensuing decade, every "missing" Federal dollar in transit support was replaced by increased fare revenues and State and local tax dollars. As noted earlier, from 1981 to 1993, State and local transit assistance doubled in constant dollar terms.

In explaining public support for transit services, this book addresses in monetary terms the benefits that the public attributes to transit. Formal economic analysis of how people value transit offers evidence that the annual value of transit far exceeds transit's annual costs to taxpayers.

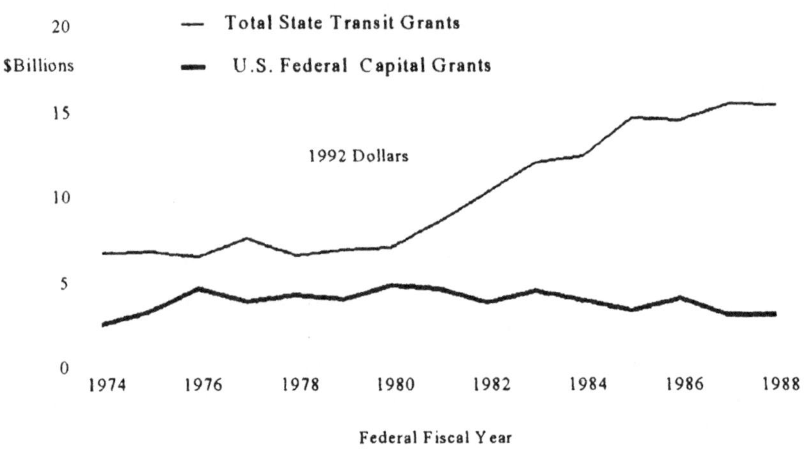

Figure 1.4 Federal and State Transit Grants, 1974 – 1988

Source: Federal Transit Administration, *Statistical Summaries* and Congressional Budget Office, *1995 Infrastructure Study.*

Transit Service as a Public Policy Function

The concept of "public policy functions" is reserved for enduring and widespread public sector operations such as public schools, law enforcement, certain utilities, and public transit services. Transit services appear to have earned a place among a select group of core public operations for which society, through annual local and State budgetary processes, maintains enduring and widespread financial support.

In most communities, transit's only function is to provide low-cost mobility to the young, the elderly, people with disabilities, and people who cannot afford their own cars. As will be seen in Chapter 4, this function—let us call it "affordable mobility"—has much greater economic value than is recognized by analysts generally. Low-cost mobility exhibits extraordinary value to passengers simply because they use transit for their most essential travel, and least for discretionary trips.

In large central cities and in sections of many smaller areas, transit serves a second function. Transit services that are a short walk from homes and business, and which require very little waiting, can support livable neighborhoods by enabling households, businesses, and institutions to avoid the costs of owning, operating, and parking automobiles. Students and their parents save in expenses and tuition when college campuses are designed for bicycles, wheelchairs, pedestrians, and buses instead of cars, parking lots, and highways. Commercial centers in such cities as New York, Chicago, and San Francisco achieve efficiencies in the concentration of hundreds of thousands of workers, business exchanges, and support services in skyscrapers clustered around stock markets, financial markets, wholesale garment districts, world trade centers, city halls, entertainment "capitals", and commodity exchanges. Households achieve large monthly savings in the relatively fixed costs of auto ownership when they reside in walking distance of transit stations which attract a mixture of retail businesses, public services, and social activities (mixed use development). Relying on hedonic research techniques, the value of livable community benefits to residential areas is examined in Chapter 5.

In numerous severely congested travel corridors across the United States, transit serves a third public policy functions, namely the measurable alleviation of highway traffic congestion. By enabling would-be motorists to bypass congested highways via a subway, a reserved HOV lane, or other high-speed transit facility, transit maintains a stable travel time equilibrium for transit passengers and motorists alike. This intermodal process produces immediate time savings for all commuters in the corridor, helps avoid the costs of highway construction, and reduces the undesirable side effects of traffic congestion, such as air pollution and urban blight. Results of econometric analyses and corridor measurements bearing upon transit's role in congestion management are presented in Chapter 3.

The Transit Policy Debate

A primary goal of the research reported in this book is to engage the national policy debate over transit's role in American society. As discussed earlier, however, much of the policy debate in the United States has been led by professionals making arguments that are literally academic. The arguments have relied on a dissection of public policy goals in "technical" terms, characterized earlier as synoptic or "modeling" analyses. Instead of efforts to measure the monetary value of bundled transit outcomes to households and taxpayers, the technical collegium has persistently measured patronage, costs, mode split, and travel times. Alternatively, some researchers have compiled taxonomies of transit benefits, listing every conceivable intended or unintended consequence of transit: physical, economical or spiritual.[45] The transit goals so constructed denature the decision making process, substituting a fanciful "rational" decision making process with no counterpart in decision making. Colored in this way by largely unexamined assumptions about decision making (e.g., budgets correspond to opinion polls), the academic debate has had a largely detrimental influence on transit investments in the United States.

As maintained above, the academic debate is largely beside the point. Even if transit were to achieve great gains in cleaner air, safer highways, more efficient urban development, and quicker journeys to work, an analysis reporting on these virtues alone would carry limited value for setting transit budgets in the public sector. Legislative bodies acknowledge "intended outcomes" of public policies. But these outcomes only have budgetary heft to the extent they have measured value to constituencies. Indeed, when monetary value to constituencies is not measured, other means of measuring value prevail, such as charity, vocalism and informal influence. For example, over the years, very few have quarreled with the transit goal of providing basic mobility for people who do not drive cars. Lacking a measure of their economic value, however, transit services for this market niche tend to be set according to what a largely disenfranchised constituency will tolerate. As a result, affordable mobility transit services in the United States tend to be "politically correct" without being as effective as they could be.

With these considerations in mind, we contend that the genuine policy debate belongs at the bus stop, in the transit garage, and at the local transit policy board. Imperfectly, this grounded policy debate has reached legislative bodies in the form of policy goals that are more substantial than generally acknowledged in the literature. The three functions of transit are not new to transportation policy debates, just eclipsed in the academic literature. They were articulated in 1962 by Secretary of Commerce Luther H. Hodges and Housing and Home Finance Administrator Robert C. Weaver when they proposed enactment of Federal transit support:

> Increased emphasis on mass transportation is needed because only a balanced system can provide for:

a. The achievement of *land-use patterns* which contributed to the economic, physical, and social well being of urban areas;

b. The *independent mobility of individuals* in those substantial segments of the urban population unable to command direct use of automobiles;

c. The improvement in *overall traffic flow and time of travel* within the urban areas; and

d. Desirable standards of transportation at least total cost [our emphasis].[46]

The beginnings of Federal transit support had a very strong planning emphasis because transportation in general was recognized as a decisive influence on highway traffic, basic mobility, and land use patterns. Only with effective coordination was there optimism for new transportation investment to be effective in combating traffic congestion. Planning could help preserve and enhance neighborhoods disrupted by necessary transportation investments in their midst. Planning would incorporate the mobility needs of disadvantaged groups into transportation investment programs. Left to markets and *ad hoc* political deals, development would only erode the urban landscape.

The early hopes for planning have not been realized. Instead, the planning process became a venue for convening political jurisdictions having stakes in transportation projects. Instead of intermodal coordination, transit, railroads, aviation, and highway agencies ignored each other. Instead of fostering coherence and access, planning bodies for the most part facilitated urban fragmentation while still denying access to disadvantaged groups and aggrieved neighborhoods. Excluded groups turned to the courts in the ensuing decades to achieve the access and coherence envisioned by Hodges and Weaver in 1962.

Significant Congressional action responded intermittently to the many local conflicts incited by major transportation projects. For example, in 1973 Congress empowered local areas to "trade in" "interstate transfer" highway trust fund dollars for transit funding from general revenues. Federal aid highway dollars could be converted to transit grant purposes—with a higher local share. In 1982 Congress set up a transit account in the Highway Trust Fund. The Intermodal Surface Transportation Efficiency Act of 1991 (ISTEA) consolidated and broadened these gains.

ISTEA was a second run at bringing access and rationality to bear on local transportation investments. A model of Federal devolution, ISTEA lowered barriers to allow local authorities to use transit funds for highways and vice versa, depending on local needs analysis. ISTEA gave Metropolitan planning agencies new authority to influence State transportation plans. ISTEA opened the metropolitan planning process to "new partners" in transportation, including grass-roots associations.

Since enactment of Federal transit legislation in 1964, transit has been viewed narrowly as a failing industry overtaken by economic and demographic forces. In the prevailing mood, reinforced after 1981 with no net growth in financial support, transit was considered to have little influence on traffic congestion in its immediate environs. With explosive growth in suburban residence and jobs, transit was considered a follower of land use trends, and hardly a force capable of exerting influence on land use patterns. The mobility transit afforded disadvantaged groups was considered an "inferior good" that would be abandoned as soon as people could afford a car. The only goal that seemed to have merit was to control transit operating costs.[47] Stung, perhaps, by transit's perceived failure to attain the loftier goals set for it, efficiency became transit's paramount goal in the early 1980s. This framework has been changed very little since adoption of ISTEA in 1991.

In principle, the three public policy functions of transit in the United States often are mutually supportive. High density transit systems offer superlative affordable mobility while simultaneously supporting valuable residential and commercial areas. Less car ownership in livable communities means less traffic congestion. A significant share of commuters bypassing congested highways cannot afford cars, and are therefore not potential motorists. Instead, they are riding transit to save money rather than time, i.e., low-cost mobility. The overlap of functions is considerable.

With decades of chronic underfunding in the name of efficiency, however, transit's multiple functions have come into conflict. Transit's greatest deficits are generated by congestion management peak-hour services to distant suburban residential areas. The shift of transit resources to congestion management has compelled transit managers to compromise services on relatively unpeaked crosstown, local, and radial routes that support low income and high density neighborhoods. Because these less peaked services tend to produce comparatively high fare revenues compared to costs, the shift of services to the commuter routes has driven up overall deficits. This is discussed more fully in Chapter 2.

Neglect of transit infrastructure has followed the shift to commuter services and, together with poorer overall services, has eventually contributed to the central city blight that continues to drive middle income households to the suburbs. Where central cities have remained employment centers, the budgetary demand for commuter services has increased with suburban sprawl. Transit's dependence on suburban tax dollars has grown with the size of the transit deficit, requiring more and more suburban commuter services and the cycle repeated.[48]

As a consequence of this cycle, during the 1970s, daily transit journeys-to-work within central cities declined by 533,000 to 3.267 million while suburban commutes to central city jobs on transit increased by 403,000 to 1.191 million.[49] Since the transit commuter flow from suburban residences to central city jobs was stable from 1980 to 1990, it appears that this pattern was arrested after 1980.[50]

This "interchange" between central city and suburban transit services is reflected in the transit journey to work data presented in Figure 1.5. Unfortunately, urbanized area data were not available for 1990, so that it is necessary to compare the 1970s based on Urbanized Area (Urban) data with the 1980s based on Metropolitan Area (MSA) data. Nevertheless, the pattern described above is clear, decline of central city-based commutes versus increased suburban-based commutes.

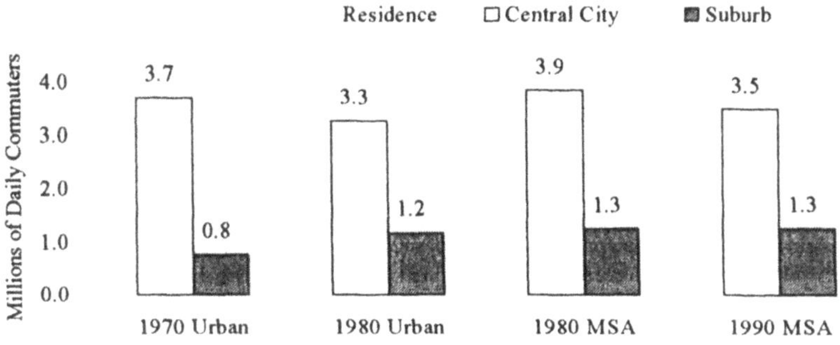

Figure 1.5 Daily Journeys to Central City Jobs

Source: United States Department of Commerce, United States Census Journey to Work Files, 1970, 1980, 1990.

In public policy arenas like transportation, where taxes, subsidies, and regulation intervene between market supply and demand, it is extremely important to take careful measurement of value. Heretofore, in a policy environment dominated by the construction of transportation infrastructure, policymakers have been impatient with difficult measurement. The public has paid a price in uncertainty, dead end projects, litigation by affected communities, and other unwelcome surprises. A sea change is underway in which transportation economists and planners are studying the costs and benefits more closely, particularly the value of time. In so doing, we are discovering the new territory exemplified in the pages to follow.

Chapter 2 of this book examines in more detail the idea that transit services are directed by local public policy to address three market niches in which automobile performance is stifled, respectively, by (1) natural barriers to auto ownership including initial cost; (2) peak period congestion on urban arterials, and (3) the superiority of foot travel and mass transit in high density residential neighborhoods and commercial centers. With the substitution effect of transit on auto use as its focus, and with avoided auto ownership costs for rudimentary valuation, Chapter 2 explores transit's benefits not only to transit customers. It explores the proximate indirect benefits of transit use to specific local constituencies which appear to have accepted the establishment of transit as a municipal function. Chapter 2 concludes with an effort to valuate the spillover impacts of transit substitution on clean air.

The succeeding three chapters take up formal economic analysis of transit's major benefits to passengers and its constituencies. These chapters roughly correspond to the three public policy functions discussed in Chapter 2, but the analyses they contain are quite independent of this public policy framework. Each is a free standing analysis. Chapter 3 explores the dynamic equilibrium which exists in a number of severely congested commuting corridors served by rapid rail and high occupancy vehicle lanes. Chapter 3 also conducts an econometric analysis of the travel time savings that result from the operation of that dynamic equilibrium. Chapter 4 is an econometric analysis of the low-cost mobility benefits that transit offers to low income households. Chapter 5 begins with an historical discussion of transit in relation to the evolution of American cities and suburbs. Chapter 5 then reports the results of hedonic analyses of transit oriented neighborhoods, showing the effect of transit and transit-related density on real estate values.

The concluding chapter draws obvious lessons from the research, the most significant of which is that the benefits to passengers and nonpassengers exceed the $22 billion passengers and taxpayers spent in transit in 1993. Moreover, because of the implicit "devaluation" of transit benefits owing to poor measurement, transit's potential benefits exceed the most ambitious policy prescriptions. With an intuitive understanding of transit's ability to substitute for cars in congested travel corridors, voters have supported the construction of a new generation of rapid transit systems across the United States. But the public conversation is only beginning concerning transit's ability to make our metropolitan areas better by increasing the walkability of neighborhoods and commercial districts. For low income Americans who depend on transit systems, transit offers only miserly services to the four corners of the metropolitan areas where the modern United States economy creates jobs and wealth.

Notes

1 The Federally estimated capital costs to sustain transit services in 1995 were $7.9 billion; the costs to improve services were $12.9 billion; and the actual capital investments were $5.6 billion. This ratio of transit opportunities and funding is chronic.

2 Summarized in Table 2.13, Page 45 of Chapter 2.

3 Polyarchy, "rule by the many", is used by political scientists in preference to "democracy" which connotes many practices not uniformly present in actual democratic institutions. Cf., Charles E. Lindblom, *Politics and Markets*, (New York: Basic Books, 1977), p. 132.

4 The seminal article is, Buchanan, James, "Social Choice, Democracy, and Free Markets", *Journal of Political Economy*, 62(2) April 1954a.

5 The most influential analysis of transit expressly focused on transit's congestion management role, an understandable focus that was passed on to many subsequent studies. J.R. Meyer, J.F. Kain, M. Wohl, *The Urban Transportation Problem*, (Cambridge: Harvard University Press, 1965), p. 5.

6 T.R. Lakshmanan, Peter Nijkamp, and Erik Verhoef, "Full Benefits and Costs of Transportation: Review and Prospects", *The Full Costs and Benefits of Transportation,* (New York: Springer, 1997), p. 391.

7 Ulric Blum, "Benefits and External Benefits of Transport: A Spatial View", *The Full Costs and Benefits of Transportation*, (New York: Springer, 1997), p. 219.

8 Federal Transit Act of 1964 as amended, Section 2.

9 Cf., David Jones, *Urban Transit Policy: An Economic and Political History*, (Englewood Cliffs, N.J., 1985).

10 Arguably, without labor protections, for example, there would be no Federal transit legislation. Similar things may be said of Buy America, access for people with disabilities, and protections for private bus companies.

11 Herbert A. Simon, *Administrative Behavior,* Third Edition, (New York: Free Press, 1976), p. xxviii.

12 David Braybrooke and Charles E. Lindblom, *Strategy for Decision*, (New York: Free Press, 1961). The book elaborates on Lindblom's short article: "The Science of Muddling Through", *Public Administration Review*, Vol. 2, (1959), pp. 79-88. Lindblom's work is grounded in a large body of methodological, philosophical, and empirical literature dating from the Utilitarian philosophers Jeremy Bentham and John Stuart Mill. Lindblom invokes John Dewey's application of Pragmatism to public policy and Karl Popper's criticism of ideology, studies of bureaucracy by Herbert Simon and James March, studies of urban politics by Martin Myerson and Edward C. Banfield, and Lindblom's own analytical work with Robert Dahl. The Public Choice paradigm carries this thread in formal economic theory. The literature argues, sometimes heatedly and always convincingly, against the notion that the seemingly irrational, chaotic, and wasteful decisionmaking observed in pluralist societies is naturally less efficient or less informed than the use of economic and engineering planning models to decide public policy options. In the larger historical social science literature, Braybrooke and Lindblom represent an antithesis to command and control economies.

13 "Mr. Lindblom dislikes a *doctrinaire* attitude toward anything. He incessantly encourages the pragmatic approach to economics. It naturally follows than any reliance on absolutes, or any reference to indefeasible 'rights' is unwarranted and anachronistic." William F. Buckley, Jr., cited in Buckley's *The Lexicon*, (New York: Harcourt, 1996), p. 46.

14 Rather than reading the regulations, the student should consult an Environmental Impact Statement for a project in an area he or she is familiar with.

15 Braybrooke and Lindblom, op. cit., p. 86.

16 Patronage losses, therefore, create a budget crunch for transit boards, requiring new budget decisions. This perennial transit financial crisis, perhaps because it so unsettles incremental decision making practices, is often confused with mission failure—the perennial bureaucratic mission being that of presenting the fewest possible new requests to legislative bodies.

17 To use Simon's phrase, "bounded rationality".

18 United States Department of Transportation, 1995 Nationwide Personal Transportation Survey.

19 According to Martin Mogridge's theory of dynamic travel time equilibrium among modes. Martin J.H. Mogridge, *Travel In Towns*, (London, MacMillan, 1990).

20 Including 8 years during which the Federal government campaigned for the privatization of local transit services, saying that private transit companies would be more efficient.

21 Eighteen percent of costs in areas under 200,000 population, National Transit Database.

22 Alan Altshuler, et al, op. cit., p. 33.

23 Donald H. Pickrell, "A Desire Named Streetcar", *Journal of the American Planning Association*, Vol. 58, No. 2 (Spring 1992), p. 169.

24 Ibid., p. 159.

25 Meyer and Gomez-Ibanez, op. cit., p. 14.

26 Ibid., p. 13.

27 Alan Altshuler, et al, op. cit., p. ix.

28 A suggestion consistent with Don Pickrell's findings.

29 Draft paper for Federal Transit Administration, prepared by Cambridge Systematics, September 1988.

30 Over 2,000 pages of reports, hearings and testimony in: United States House of Representatives, *Transit Development Program for the National Capital Region, Report No. 1005*, (Washington, D.C.: United States Government Printing Office,1963); United States House of Representatives, *Transit Program for the National Capital Region*, Hearings, (Washington, D.C.: United States Government Printing Office,1963); United States House of Representatives, *Urban Mass Transportation Act of 1962*, Hearings, (Washington, D.C.: United States Government Printing Office,1962); United States House of Representatives, *Urban Mass Transportation Act of 1963,* Hearings, (Washington, D.C.: United States Government Printing Office,1963).

31 David Jones, op. cit., (Englewood Cliffs, N.J., 1985), p. 6.

32 Ibid., p. 124.

33 Transit Cooperative Research Program, "The Impact of Civil Rights Litigation Under Title VI and Related Laws on Transit Decision Making", *Legal Research Digest*, Number 7, (Washington, D.C.: Transportation Research Board), June 1997.

34 Ibid., p. 18.

35 Ibid., p. 19.

36 Ibid., pp. 19-20.

37 "Job creation" by public spending is usually an economic "transfer" from another sector of the economy from which the revenues are drawn, and is not as such a benefit to society.

38 As early as 1938, Federal funds were used to for subway construction in Chicago and New York City. Brian J. Cudahy, *Cash, Tokens, and Transfers: A History of Urban Mass Transit in North America*, (New York: Fordham, 1982), pp.176-177.

39 Braybrooke and Lindblom, op. cit.,

40 James M. Buchanan, "Social Choice, Democracy, and Free Markets", *Journal of Political Economy,* April, 1954 (as cited in Amartya Sen, "Rationality and Social Choice", *The American Economic Review,* (March 1995), p. 3.

41 Ibid., p. 18.

42 Charles E. Lindblom, *Inquiry and Change,* (New Haven: Yale, 1992).

43 Arrow, Kenneth J., *Social Choice and Individual Values*, 2nd Edition, (New York: Wiley, 1963).

44 The seminal article is, Buchanan, James, "Social Choice, Democracy, and Free Markets", *Journal of Political Economy*, 62(2) April 1954a.

45 Edward Beimborn and Alan Horowitz, *Measurement of Transit Benefits,* Center for Urban Transportation Studies, University of Wisconsin-Milwaukee, (1993).

46 Hon. Luther H. Hodges, "Urban Transportation–Joint Report to the President by the Secretary of Commerce and the Housing and Home finance Administrator", *Hearings on the Urban Mass Transportation Act of 1962* (H.R. 11158), House Committee on Banking and Currency, (Washington, D.C.), p. 38. These goals are hardly

consistent with a temporary program to put transit on its feet. Rather, they imply an effort as enduring as the Interstate highway program.

47 Altshuler, *op. cit.*, pp. 44-49.

48 A conclusion reached by many students of transit, cf., John Meyer and Jose Gomez-Ibanez, Autos, *Transit and Cities*, (Cambridge: Harvard, 1981), p. 55.

49 Urban Mass Transportation Administration, *Transit Performance and Needs Report*, 1987.

50 United States Census Data File prepared for FTA and the American Association of the State Highway and Transportation Officials (AASHTO), with the assistance of Alan Pisarski.

2 The Public Policy Functions of Transit Services in the United States

Automobiles and Transit Services

In the last half of the 20th century, automobile travel has become the norm in the United States, while transit patronage has languished. In 1990, as shown in Figure 2.1, American households on average owned 1.4 private

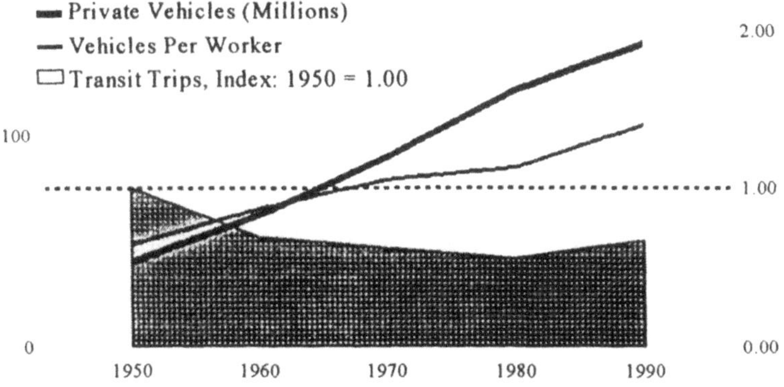

Figure 2.1 Auto Ownership and Transit Patronage in the United States, 1950 – 1990

Source: American Automobile Manufacturers Assocation, *Motor Vehicle Facts and Figures*, 1993 and Federal Transit Administration, *National Transit Database*.

motor vehicles for each person who worked outside the home, compared to 0.64 cars per worker in 1950.[51] From 1950 to 1990 transit patronage declined 30 percent. In 1993, United States households spent nearly 408 billion dollars to own and operate automobiles.[52] Total transit expenditures in the same year were approximately $22 billion.

The real consumer *cost* of auto ownership and use has declined steadily, even since the 1970s when auto manufacturers have been required to increase passenger safety, fuel efficiency, and reduce harmful vehicle emissions. Figure 2.2 compares the declining cost of auto travel with increasing transit fares during the 1980s and early 1990s. The decline in bus patronage corresponded to increasing fares and declining auto costs. Public policy has responded with expanded rapid and commuter rail transit services that compete directly with autos stifled by traffic congestion.

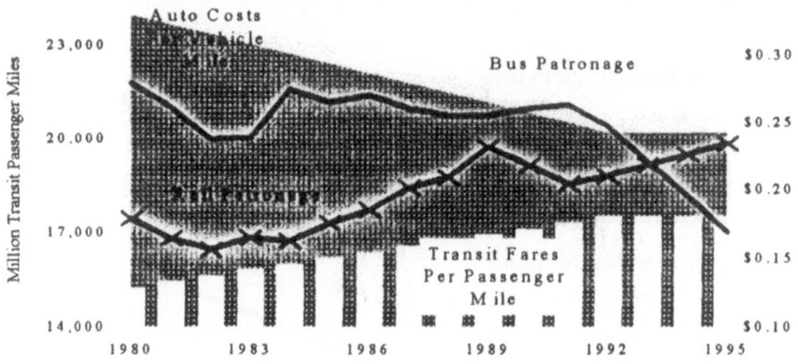

Figure 2.2 Auto Costs and Transit Fares, 1980 – 1995

Source: United States Department of Commerce, *Statistical Abstract of the United States* and Federal Transit Administration, *National Transit Database.*

The impacts of automobiles on household economics cannot be overstated. While the purchase of a motor vehicle is a financial shock for many households, a car is usually a sound capital investment. The annual cost to own and operate the average automobile in 1993 was $2,714.[53] The annual cost to operate a new car in 1994 was $4,665. Most of these costs are the fixed costs of ownership. Once purchased and insured, each trip with the vehicle is remarkably cheap. Of the $2.71 Americans paid to drive their cars ten miles in 1993, on average, only 19 percent or 52 cents was for gas and oil. The remaining $2.19 was the purchase price, interest on the loan, insurance, maintenance, and other ownership expenses. Americans in 1993 paid $6.02 daily for each car parked in their driveway. A round-trip to the grocery two miles away added only 21 cents.

The disparity between the large fixed costs of ownership and the small incremental costs of each auto trip is the automobile's greatest virtue for the household. The broad appeal of convenient, personal, exclusive and cheap travel in air conditioned comfort while listening to one's favorite music is obvious. Many cars are machines of beauty in art, design, engineering, and performance, rewarding their owners with immediate as well as enduring satisfaction.[54]

Transit economics are quite different. Even with heavily subsidized transit services, the out-of-pocket bus fare usually exceeds the cost of gas for a comparable auto trip. The transit passenger also walks to a bus stop or rail station, waits for a vehicle, and often stands for part of the trip. Transit services do not reach many areas, even within highly developed metropolitan areas.

With fixed revenues since 1982, the only way transit could compete in most market niches was to shift transit resources from bus services to higher quality rail services. The resulting trends in patronage for bus and rail are shown in Figure 2.2. While not exactly a "losing" market battle for transit, Figure 2.2 illuminates a recurring transit dilemma among transit's local policy goals. This dilemma is addressed later in this chapter and in Chapter 6.

To improve their quality of life, American households find they can substitute cheap and pleasurable auto trips for many other budget items. We travel to ever larger retail stores for cheaper products, at greater distances from home, to substitute cheap auto miles for higher retail prices closer to home. We buy larger homes in more pristine suburban settings, substituting cheap auto miles for higher priced real estate that would be closer to work. We even shop by car for schools, day camps, churches, and sports leagues that are best suited to our needs. Across the spectrum of household needs, cheap auto travel affords consumers countless opportunities to shop for bargains and thus drive down living costs. These household efficiencies accrue to the general economy. Thanks to the automobile, we have come to consider our daily sphere of economic activity to encompass a large segment of the metropolitan region. As a consequence, the metropolitan region has come to be the standard unit of economic activity.

The ability to substitute cheap auto travel is conferred on nearly anyone who owns a motor vehicle. Conversely, households without cars are denied many of the economies of the modern American household. How does a parent shuttle pre-teen children to soccer games without a car? How do parents attend school events after work without a car? How does a parent make weekly supermarket run without a car? How is a teenager to bring her or his date to the senior prom without a car? Could I earn more if I had a car to look for a better job? What fraction of the opportunities in metropolitan America are available to workers without cars? Cheap auto travel has become so central to the economy of the household, to leisure time, to so many facets of daily life, it is puzzling how a household without a car can belong to the mainstream of American economic life. Even most households with below poverty incomes are willing to purchase their own motor vehicles for the obvious benefits.

Generally, the inherent merits of automobile ownership are overwhelming. But significant travel circumstances—travel market niches—exist in which private vehicles are clearly inferior to the alternatives. The most obvious example is commercial travel between metropolitan areas that are, by car, more than two hours apart. This market niche is filled by commercial airlines, Amtrak, intercity buses, and rental cars. Few people enjoy driving more than a few hours at a time, whatever the comfort and luxury of their vehicle. So we fly. We rent cars for temporary use. Taxis are perfect for the occasional trip across an unfamiliar town. Walking, jogging, bicycling, and skating even contribute to mobility (and health) in important segments of the economy. Some say the only way to see the United States is on the back of a motorcycle or a horse. Many prefer an accessible charter bus. Parking a car can be a chore.

In urban America, public transit serves three market niches that are not adroitly served by private autos and other travel modes.

> First, in nearly every urban area, transit serves a *basic mobility* function for children, elderly people, people with disabilities who are unable to drive, people who cannot afford their own cars, and motorists whose car is in the shop.
>
> Secondly, in certain urban areas, rapid transit enables a critical number of commuters to bypass severely congested freeways and thus *save travel time* for themselves and motorists alike.
>
> Third, in a number of commercial centers, urban neighborhoods, retirement communities, and towns with large college campuses, transit facilitates a *pedestrian friendly* streetscape in which walking, elevators and bicycling are more common than driving.

To casual observers who seldom board a bus, transit often is identified with disadvantaged people, some living at the margins of the economy. This is largely true, and is discussed below. However, a much broader cross-section of American society is served directly and indirectly by transit. In truth, even as the automobile pervades our lives, we still substitute transit for auto travel and thereby reduce the expense of auto ownership. In so doing, our use of transit generates benefits to specific other groups and to the public at-large. These indirect benefits win the enduring financial support of taxpayers in municipal, county, and State budgets throughout the United States.

Public Support for Transit Services: A Conceptual Framework

It is not the purpose of this book to *fully* explain public financial support for transit. It is to contribute to such an explanation by identifying benefits to three distinct "publics" that influence in the long term the annual budget processes that have provided public funds to transit in several hundred urban areas for the last quarter century. Public opinion that is attentive to transit services and the budgets that support them could be classified in a number of different ways, such as by income, or by property ownership, residential location, proximity to transit, etc. However, for the purposes of this Chapter, we distinguish transit's publics according to the economic mechanism by which the benefit is conveyed to them.

Transit supporters with a financial interest in transit expenditures per se are omitted from this analysis. Although they are important elements of the budgetary process, organized labor, contractors, and others with pecuniary interests offer little explanatory power for budget priorities in the long term. Analytically, political support for income transfers through tax collections and public expenditures are a constant factor in the decision making process and thus offer little power for explaining support for one budget item versus another over time. Also, job creation and other "pecuniary" benefits of transportation expenditures constitute benefit transfers rather than net social benefits.[55] In his 1979 book, Alan Altshuler explained his exclusion of such "pork barrel criteria" as follows:

> "We explicitly omitted pork barrel criteria (such as the volume of jobs and contracts generated by transportation expenditures) from the list [of criteria for evaluating urban transportation strategies], on two grounds. First, such effects bear only an incidental relationship to the urban transportation system. Second, the inclusion of pork barrel effects as genuine benefits will almost always lead to the conclusion that public expenditures should be increased—since, by definition, they yield their worth in payrolls and profits, in addition to whatever other benefits they produce".[56]

The financial benefits of transit are real and legitimate, but as object of analysis they are "private" in the sense that they accrue to persons in their household or occupational roles rather than as citizens or taxpayers. They are also private in the sense they have no *public* policy merit or relevance. Private beneficiaries are not attentive "publics".

Transit's financial support comes from three strata of beneficiaries, divided by classical public choice categories, as arrayed in Table 2.1. Transit passenger households receive direct benefits from their use of transit in exchange for fare payments. Local taxpayers, organized in political jurisdictions, "subscribe" to bundled transit servicees (bus stops) in exchange for regionwide sharing of the subsidy burden. General taxpayers, organized by States and Federal agencies, are willing to pay for benefits that spill over to the general public from transit services provided locally (e.g., reduced vehicle miles per capita).

Table 2.1 Public Choice Concepts for Transit Benefits

		Public Choice Categories		
Concept		Markets	"Clubs"	Public Goods
Transaction		Fare Purchase	Bargaining	Majority Rule
Distribution		Trips	Service Areas	Spill-Over
Beneficiary	e.g.	Passengers	Motorists	Californians
Conditions	e.g.	Parking Costs	Congestion	Air Quality

The first of transit's attentive "publics" comprises transit riders whose benefit is proportional to the transit services they use. In general, transit riders are obvious transit supporters because of direct benefits they receive as subsidized customers. One way to measure this benefit is merely to count the fares they pay. This is a commonly used measure of benefit, and it is reinforced by well established data on price elasticities, which measure the general effect of price and services change on the demand for services. However, the markets for most transit services are highly distorted by implicit subsidies to alternate modes, transit fare subsidies, and other public sector interventions. Thus, fare alone is a misleading measure of the value households receive when they rely on transit. Chapter 4 of this book treats "consumer surplus" as a measure of the benefit of transit subsidies to low income households in particular.

In many household budgets, transit produces a true substitution effect. This is best measured by rates of vehicle ownership. Below expected vehicle ownership is not only the strongest "predictor" of transit use, the financial windfall from avoided vehicle ownership is a reasonable measure of transit's impact on the household's transportation budget. Above all, vehicle ownership is a promising approach to determining the monetary value of substitution.

"Avoided vehicle ownership" is an imperfect measure of household benefit. It will understate savings to the extent that it ignores annual travel time savings. On the other hand, vehicle ownership savings may overstate benefits a portion of which is shifted by the real estate market to higher property values and rents A more elaborate model is needed to sort out the net benefit to households. However, avoided vehicle ownership is an excellent starting point comprehensively to account for the benefits that households realize when their members patronize transit services.

Two other "tiers" of "attentive publics" support transit budgets; people who have a conscious stake in transit though they may never use it. The first of these are called "constituents" whose benefits from transit are contingent. The transit benefit comes from the constituent's proximity to transit stations, from the constituent's use of congested highways that are affected by transit, or from independent mobility that transit affords a relative, a friend, a colleague, or a fellow citizen.

Transit's local "constituents" are people who may seldom use transit themselves, but who nevertheless readily can see (or presuppose) a difference that transit makes in their lives, a difference that translates into discernible political support for transit budgets. Transit's "constituent" benefits are exclusive benefits in that they can be denied to neighborhoods and political jurisdictions that do not provide financial support.[57] The cultivation of constituencies so defined has been the fulcrum of transit's change from a private to a public service since the 1960s—and transit's highly regulated status since its origins.[58] The constituency for transit is reflected in the multi-jurisdictional scope of most transit services in the United States, spelled out in multi-jurisdictional financial charters. Owing to the exclusive yet grouped nature of these benefits, economists have called them "club" benefits.[59]

The second tier of transit's "attentive public" consists of the public at-large. The public looks to transit for nondivisible public goods such as cleaner air, lower gas prices, lower highway budgets, less national product devoted to transportation costs, and other diffuse benefits that transit provides equally to all members of society. Transit's diffuse benefits depend on its efficiency in substituting for automobile vehicle miles (VMTs). These are called "spillover" benefits.

Transit's spillover benefits enter the transit budget most significantly at the national level. Like most financial programs at all levels of government, Federal transit funding is fought for and won by "special interests", including transit managers, transit equipment suppliers, building contractors, and consultants. Sometimes these interests are not as narrow as the term "special interest" connotes. But transit budgets are also supported nationally by Mayors, Governors, and other State, local and regional "public interest" groups. These interest groups invoke spillover transit benefits in behalf of Nationally established goals for clean air, cities, employment, and quality of life.

Thus, transit's spillover benefits help to legitimate Federal support for transit, perhaps as a "fig leaf" for parochial concerns. In so doing, however, these benefits have won over National constituencies for national goals. In this way have National constituents adopted transit and broadened its mission, often subordinating more parochial transit interests in the process.

For example, recipients of Federal transit funds are required to satisfy Federal civil rights goals in service delivery and accessible vehicle design. Transit buses have been required to serve as a test bed for new clean air engine technologies.

Public sector rhetoric often mimics the language of markets, for example, citing increased market share of commuters as a goal for transit. But the public seldom engages in market behavior in the usual sense of numerous buyers and sellers in an unregulated free market. Rather, established public programs such as transit are more usefully seen as fulfilling "public policy functions". Transit is part of the "infrastructure" function of government or the "transportation" function. The original Federal home of transit programs was the United States Department of Commerce where they performed a commerce function. Transit programs later moved to the Department of Housing and Urban Development where they pursued urban development functions. "Function" is an ambiguous term.

As used below, the term "public policy function" or transit "function" is not meant in the Departmental or budgetary senses. Rather, "public policy function" refers to that aspect of the transit system corresponding to a field of public interest. Transit's three public policy functions correspond to transit's three "market" niches, not by accident, but by convergence inherent in local budgetary processes. Acting independently over time, transit budgets across the country have converged in establishing three objectives for transit: local cost mobility in most areas, congestion management in severely congested commuter corridors, and livable neighborhoods in densely developed sections of many cities and towns.[60] In diverse local areas, constituencies for these functions have evolved and have exerted influence over time.

Transit serves the functions by substituting for private vehicles. To say that private vehicle travel is the norm in the United States economy is to say that households without cars are economic anomalies which tend to substitute an alternative mode of transportation. Quantified variance in household vehicle ownership thus is a reasonable measure of transit services fulfilling public policy functions as a substitute for auto travel.

Transit and Vehicle Ownership: The Household Benefits

In 1990, American households on the whole owned 0.7 vehicles per person. Generally, vehicles-per-person declines with household size, starting at more than one vehicle in one person households, and reaching 0.5 vehicles per person in five-person households.[61] As shown in Table 2.2, for households located within six blocks of transit, the number of vehicles per person is considerably less than the United States norm.[62] The average for households above poverty without transit access was 0.74 vehicles per person. Households located less than six blocks from transit, however, owned 0.66 vehicles per person in 1990.

For households with income below poverty, the average was 0.58 vehicles per person. Households above poverty had 0.16 or 27 percent more vehicles per person than households near or below poverty. Low income households located near transit had 0.4 vehicles per person, nearly one-third fewer than low income households without transit.

This relationship between auto ownership and transit proximity strongly supports the idea that the household savings associated with transit as a substitute for auto ownership is a valid measure of economic value.

As mentioned earlier, transit services offer a competitive substitute for private vehicles in three market niches:

1. Minimal transit services make possible a basic level of mobility for children, elderly people, households which cannot afford their own vehicles, people who cannot drive due to physical disabilities, and people who use transit when their car is temporarily unavailable.
2. Rapid transit operations (which operate on their own tracks or busways), High Occupancy Vehicle (HOV) and High Occupancy Tollway (HOT)[63] facilities provide a means to bypass congested

highway traffic for commuters and others who must travel during rush hours.

3. Concentrated transit services preserve and foster the development of neighborhoods, campuses, and commercial centers that can rely on walking (and elevators) as well as transit to meet the majority of daily travel needs.

Table 2.2 Transit Access and Household Vehicle Ownership, 1990 Per Person Vehicle Count

	Nearest Transit to Residence		
	< 6 blocks	6 blocks – 2 Miles	No Transit
Households Above Poverty*	34.7	43.0	31.6
Household Size	Vehicles per Person		
One Person	0.89	1.04	1.04
Two Persons	0.79	0.92	0.91
Three Persons	0.64	0.77	0.77
Four Persons	0.53	0.60	0.61
Five Persons	0.45	0.50	0.49
1- 5 Person Above Poverty Households	0.66	0.73	0.74
Households Below Poverty*	7.3	0.7	6.0
1-5 Person Below Poverty Households	0.40	0.57	0.58

*Millions

Source: Author's Analysis of 1990 Nationwide Personal Transportation Survey.

Basic Mobility: Low Cost Access to the Regional Economy

The "entry fees" of auto ownership (e.g., down payment, insurance, interest, and taxes) form a significant barrier to normal mobility for people just starting out on their careers, newly arrived immigrants, and workers with very low incomes. Averaged over the typical annual 11,000 miles, autos cost $1.42 per passenger for a ten mile trip,[64] a cost that is comparable to a transit fare. But the first auto trip can cost $2,000, counting the price of a down payment, interest, insurance and other auto ownership expenses. A cheap rental car (with insurance) might be arranged for as little as $30 per day—with unlimited mileage. For most households, of course, this hurdle to auto ownership is overcome as their earnings increase.

Children below driving age, elderly people who cannot safely drive, and people with certain disabilities are also denied the extraordinary mobility that has become normal in the American economy. And although motor vehicle repair services often provide lender cars for their customers, nearly everyone is sometimes without the car for a day or two. As a result, a market niche exists for low-cost motorized mobility for millions of Americans on a continuing and on a temporary basis.

Transit users, on average, pay $1.80 for the first and last ten-mile trip. Households that are struggling economically or individuals who do not need to travel 11,000 miles per year (e.g., children, students, and retired people) find they can save a great deal of money by foregoing auto ownership and residing near transit stations and bus stops. In fact, many people choose their neighborhoods principally for transit access.[65]

Table 2.3 presents data on households that used transit for *low-cost mobility* in 1990. This group includes transit riders from approximately 5 million households.[66] The group includes licensed drivers with incomes above poverty who own automobiles, but who report using transit during *off-peak* hours of the day (34 percent). These households neither use transit to bypass peak hour traffic congestion nor to substitute for vehicle ownership. Rather, for many in such circumstances, transit is a low cost back-up mode for the car. The majority of people looking to transit for low-cost access, however, are people who depend on transit because they do not drive (no license) or report no motor vehicle in their household. These are mainly children, elderly people, and people with near or below poverty household incomes.[67]

In 1990, 2 billion linked transit trips were made because transit offered basic mobility at a low per trip cost,[68] accounting for 43.2 percent of transit's patronage. These trips, however, tended to be shorter than average, so that they only accounted for 40.5 percent of transit passenger miles. Auto ownership among this group (0.37 per person) was far below the auto ownership of the same mix of poverty status groups[69] (0.68 per person) that did not use transit. This suggests that transit access may have saved transit-dependent households, on average, as much as $2,443 per year in auto costs, for an aggregate annual saving of $9.3 billion.[70]

Table 2.3 Household Savings from Low Cost Mobility, 1990

	Per Year	United States
Linked Transit Trips[b]	2,081	43.2 percent
Passenger Miles[b]	20,408	40.5 percent
Per Person Vehicles Owned	0.37	Base = 0.68[a]
Vehicle Savings per Household	$2,443	
Aggregate Vehicle Savings[b]	$9,288	
Per Person Vehicle Miles (VMT)	3,745	7,349[a]
Per Person Vehicle Mile Savings	3,604	
Aggregate VMT Savings	40,802	

[a] For the same poverty status mix not reporting transit use.
[b]Millions

Source: Author's analysis of 1990 Nationwide Personal Transportation Survey.

In the majority of urbanized areas, the proportion of population that uses transit is minuscule.[71] Yet, this does not mean transit goes unnoticed in most areas. Rather, transport for the young, the elderly, people with disabilities, and people who simply cannot afford cars is the most universal and widely recognized service transit provides. In the many urban areas with rudimentary transit service and meager patronage, public support is probably best explained by a richly diverse, if thinly distributed, array of people who depend on transit. Nearly everyone has contact with employees, co-workers, relatives, and retail clerical staff whose mobility would be severely compromised without transit. Intense popular outcry can be provoked by reduction in even the most skeletal transit services.

Thus, often the merest transit is the most valued transit to the local community, a curious phenomenon that this book addresses in Chapter 4.

Other than rush hour commutes to suburban residences at the end of the work day (for people who might otherwise crowd arterial highways), United States transit services for suburban destinations tend to be minimal. This means that transit dependent people seeking jobs, training, shopping, and other activities in the sprawling suburbs, are effectively denied the *regional* mobility needed for full participation in the modern United States economy.[72] The effects of this economic isolation may be tolerable for many retired people, children, students, and others with special or temporary circumstances. However, members of the workforce who depend on transit are cut off. The "spatial mismatch hypothesis", a name given to the concentration of underemployed labor in central cities and labor shortages in the suburbs, has attracted the attention of policymakers.[73] The public's attention on "welfare to work" may draw a wider audience to the potential role expanded transit services can play in reducing the isolation of workers.

Facing perennial funding shortages for more than two decades, expansion of transit's role in managing congestion has generally led to the withdrawal of low cost mobility services. Alan Altshuler addressed this trade-off:

> "The primary needs of the carless poor are not for improved high speed, peak period, downtown-oriented commuter services. Their mobility deprivation applies overwhelmingly to other types of trips. It can best be ameliorated by off-peak and crosstown service improvements. Such improvements, however, typically attract very low incremental load factors and almost no automobile drivers. Rather they reduce waiting and walking times for existing low-income transit users; they afford new trip-making opportunities; they reduce the dependence of carless individuals on others for automobile 'lifts,' and they replace some burdensome walking trips. Politically, such improvements attract no support from downtown business interests; they generate no construction jobs or contracts; they do not expand the base of transit system support (typically weakest in the suburbs); they cannot plausibly be sold as instruments of congestion relief or as spurs to core area development; and they are entirely lacking in technological excitement. In short, they have neither glamour nor significant pork-barrel value; the benefits are hard to measure; and they typically come at rather high cost per trip served. . . It is scarcely surprising, then, that the great preponderance of recent transit service expansion has aimed at serving

the potential transit markets of least relevance to the problem of mobility deprivation".[74]

This topic of trade-offs among transit's functions is addressed in the concluding chapter of this book.

To Bypass Congested Motor Vehicle Routes

In large metropolitan areas, millions of daily commuters use subways, regional rapid rail, and commuter train services to bypass congested freeways and avoid downtown parking costs. Where traffic congestion would otherwise be severe and uncertain, the rapid transit bypass option erodes highway travel demand by offering equivalent and more predictable travel times. This dampening of highway travel demand has the effect of "pacing" highway travel times to transit travel time.[75] The resulting "travel time equilibrium" is not only measured experimentally, but is actually experienced by commuters who switch modes in search of quicker trips.[76]

Table 2.4 indicates the role played by travel time and speed in attracting transit passengers who wish to bypass congested commuter highways.

Table 2.4 Transit Speed and Its Congestion Management Function

Peak Period Work Trips		
Urban Areas > 1 Million Population	Trip Time 30-120 Minutes	Door-to-Door MPH
Without Transit Rail Services		
Private Vehicle	26.0	31.9
Transit Bus	41.7	16.3
With Transit Rail Services		
Private Vehicle	28.6	31.7
Transit Rail	81.0	24.4
Difference for Transit (percent)	194.2	149.4

Source: Author's analysis of 1995 Nationwide Personal Transportation Survey.

Average travel speeds are calculated for peak period motorists commuting to work in urban areas over one million population. The average *highway* commuting speed is nearly identical for these large metropolitan areas with or without rapid rail transit services, 31.7 and 31.9 miles per hour, respectively. Also, rail and nonrail areas have a similar proportion of peak period highway commuters spending more than 30 minutes on each trip, 28.6 percent and 26.0 percent, respectively. During peak commuting hours transit caters to longer trips. In "bus only" areas, 41.7 percent of peak period transit commutes take 30 minute trips or longer, compared to 26.0 percent of peak period highway commuters in the same areas. But clearly transit buses are the "inferior" mode, averaging only 16.3 miles per hour, slightly more than half the speed of private vehicles. In "bus only" cities, commuters rarely choose transit to "bypass" congested highways.

The pattern is significantly different is "rail cities", where significant numbers of commuters opt for rapid transit. In 1990 81 percent of peak period *rail* transit worktrips were 30 minutes or longer, twice the proportion in bus only cities. The average speed on rail transit was 24.4 miles per hour, 50 percent faster than transit in bus only cities of comparable size.[77]

Table 2.5 shows 1990 household data on transit passengers who had two things in common: they rode transit in the peak period and at least one vehicle was available to the household.[78] All but 6 percent of these passengers had above poverty incomes. Fifty-six percent were above poverty working age people with drivers' licenses. The data presented in Table 2.5 suggest the relative share of transit use that is explained by the desire of commuters to bypass congested freeways. The 1.5 billion trips reported in Table 2.5 (about 6.1 million one-way trips each weekday) were made in 1990 by residents of suburban areas using rapid rail, commuter rail, light rail, and buses to circumvent congested freeways during peak commuting hours. While they accounted for only 32.1 percent of transit trips, they were responsible for 40.6 percent of total transit passenger miles.

In substituting for the car during the rush hours, transit enables some suburban households to avoid or delay purchase of a second or third car. As compared to the 0.73 per person auto ownership among the household poverty status mix who did *not* report transit use, the households represented in Table 2.5 owned 0.66 vehicles per person. Accordingly, households that used transit to bypass congestion achieved modest savings in auto ownership expenses, $589 per household, resulting in aggregate saving of $1.1 billion in 1990.[79]

Unlike low cost mobility and livable neighborhood functions, the congestion management function does not depend wholly on households avoiding vehicle ownership costs as such. Rather, the desire for a reasonably timely trip in the face of congested highways and high parking costs ("generalized costs") spurs the use of rush hour rapid transit. These are distinctly *travel* benefits. These travel benefits are conservatively estimated to equal the fares paid by this segment of transit passengers.

Table 2.5 Household Savings from Congestion Management, 1990

	Annual	United States
Linked Transit Trips[b]	1.543	32.1 percent
Transit Passenger Miles[b]	20.386	40.5 percent
Annual per Person Vehicles Owned	0.66	Base = 0.73[a]
Vehicle Savings per Household	$589	
Aggregate Vehicle Savings[b]	$1,120	
Annual per Household Gen. Cost Savings	$2,432	Gen Csts = $3/Trip
Aggregate Fuel, Parking, Time Savings[b]	$4,629	
Per Person Vehicle Miles	5,635	8,907[a]
Per Person Vehicle Mile Savings	3,272	
Aggregate VMT Savings[b]	19,393	

[a]For the same poverty status mix not reporting transit use.
[b]Millions

Source: Author's analysis of 1990 Nationwide Personal Transportation Survey.

Based on revenue calculations for congestion management (reported in Table 2.9 below), the per trip fare in 1993 was $1.30, giving each household in Table 2.5 $2,432 annual savings in generalized costs. The annual household travel benefit for 1993 was $4.6 billion.

Rapid transit (defined as collective services that achieve high speeds by traveling on separate rights-of-way) thrives in areas with severely congested freeways that connect suburban areas with central city economic centers. In 1990, transit's share of journeys to work in the New York metropolitan area was 53.4 percent, Boston (31.5 percent), Chicago

(29.7 percent), San Francisco (33.5 percent), Washington, D.C. (36.6 percent) and Philadelphia (28.7 percent).[80] These high transit shares reflect the saturation of freeways in critical commuting corridors. In these corridors rapid transit is essential for effectively contending with traffic congestion for the commuter as well as for transportation planners.

Travel time studies in severely congested urban corridors demonstrate that a "dynamic equilibrium" often exists among highways, high occupancy vehicle lanes, and rapid transit services.[81] For many commuters in these travel corridors, little or no travel time advantage exists for any commuting mode. Such a pattern is the hallmark of travel time equilibrium. When travel times (or generalized costs) converge, transit investments may be the most effective way in which to lower travel times on all modes. Transit sets the pattern because, alone among the commuter's options, rapid transit services thrive on very large numbers of commuters. For little incremental cost, as compared to the cost of highway expansion, transit managers are able to increase commuter train frequencies to address crowding on rapid transit vehicles. Increased train frequencies not only increase the per hour capacity of the transit system. They also reduce waits between transit vehicles. The resulting travel time improvements induce a critical number of motorists to switch to transit, which dampens travel demand for highways and thus improves highway speeds. A new equilibrium is achieved when highway travel time reaches parity with rapid transit travel time. Subsequent crowding and slowing of either mode leads to a repeat of the cycle.[82]

In commuter corridors where the dynamic equilibrium is in full swing, transit's dampening effect on highway travel demand is evident to many passengers and motorists. They recognize that transit helps to keep traffic congestion within tolerable limits and increasingly so during the cyclical economic expansions that strain the highway networks. Since the 1970s, when taxpayer-financed buy-outs of transit companies transformed the transit industry into a public service, transit's growth has been most pronounced in combating traffic congestion. From 1970 to 1980, the number of workers commuting from suburban residential areas to central city jobs increased by 3.7 million or 55 percent. The number of these commuters using transit on a daily basis similarly increased by 408,000 or 52 percent.[83] Although the same pattern of commuting growth continued through the 1980s, insufficient funding prevented transit from continuing its growth in the suburb to city commuter niche.

The earlier trend to finance transit growth reflected public support for the intuitive idea that transit can be the best use of motor fuel and other tax revenues to help combat highway traffic congestion. Increasingly, voters recognize that increased highway capacity in the most congested travel corridors is prohibitively expensive, disruptive, at best uncertain in combating congestion and clearly detrimental in its effects on the local economy and the quality of life. This understanding among attentive publics motivated the principles behind the Intermodal Surface Transportation Efficiency Act of 1991, which directed the flexible use of Federal gas tax revenues to solve traffic congestion and other transportation problems.

As mentioned earlier, motorists driving on congested freeways recognize the impact of transit services in relieving traffic congestion. The measurement of that impact, and its economic value to motorists, are discussed in Chapter 3. In metropolitan centers throughout the United States, the convergence of public opinion on the importance of transit to congestion management is no accident. In the last quarter century metropolitan referenda have been conducted and regional institutions have been created to build new transit systems in San Francisco, Los Angeles, Atlanta, Washington, D.C., Portland (OR), Seattle, Miami, and Houston. Voters have decided to extend existing systems in San Francisco, New York, Chicago, Boston, and Philadelphia, and to modernize aging systems in New York, Boston, Philadelphia, Chicago, and San Francisco. In the majority of cases, when the public had a focused concern with traffic congestion, support for transit has prevailed. Indeed, in numerous United States cities the renaissance of rapid transit has been one of the few regionwide questions producing accord across diverse groups and jurisdictions.

Transit Oriented Neighborhoods and Commercial Centers

Transit's third market niche consists of walkable neighborhoods and commercial centers served by intensive transit services. After congestion relief, the creation and nurturing of the "sense of place" has been one of the rallying points for every regionwide transit project. Many residents and merchants in these areas may only occasionally use transit. But they are keenly aware of the effects of high quality transit on their residential and commercial property values.[84] Households in these neighborhoods own fewer cars per capita. They reap numerous benefits from residing and doing business in the high density residential and commercial concentrations that can only exist amidst intensive transit services. In these areas, walking is often much more efficient and pleasurable than driving, so that transit complements walking and, together, they support freedom from the burdens of auto ownership.

So-called "livable neighborhoods" include high density central city neighborhoods, towns with large college campuses, planned retirement communities, and central business districts in which automobile travel may be onerous and unnecessary for many trip purposes. Such areas are "location efficient" in that schools, shopping, personal services, work, residences, and other "trip generators" are within reasonable walking distance of one another. Households that are located in these places are able to forego or postpone auto ownership by choice, saving money with no loss of access. In fact, the availability and cost of parking tends to become a major household and public policy issue in many livable communities. In these circumstances, many find that vehicle ownership is simply more trouble than it is worth.

Table 2.6 presents data on transit passengers who appear to use transit by choice rather than by dependency, but not for the purpose of bypassing congested freeways. These were mostly above poverty households that owned no vehicles and in which one or more member used transit in 1990. Vehicle ownership for this group was 0.00 per person (by definition), compared to 0.74 per person for nontransit users with the same above poverty-below poverty mix. They saved $3,635 per household in auto ownership expenses in 1990, the equivalent of 1.34 vehicles per household, for a national total saving of $9.6 billion.[85]

Pedestrian-oriented neighborhoods and commercial centers form a vital part of the social and economic power of metropolitan areas. High concentrations of commerce are marketplaces for retail and wholesale trade, generating orders for goods and services throughout the metropolitan area and the more general economy. High density residential neighborhoods, served by diverse retail merchants are focal points of metropolitan identity, of labor markets—particularly for newly arrived immigrants in ports of entry to the United States, and they tend to put their stamp on copycat neighborhoods throughout the region. Many such neighborhoods are populated by wealthy households. Others are working class. The value of real estate in such neighborhoods, discussed in detail in chapter 5, reflects the many amorphous and intangible benefits that emerge in these urban neighborhoods. These benefits, in turn, sustain immediate and tangible public support for transit far beyond the transit passengers who reside in these neighborhoods.[86]

Transit's role in high density urban neighborhoods and central cities is its traditional one—high density infrastructure that spawned urban streetscapes and skylines before widespread auto ownership. This role is still very much alive. The growth of commuter-oriented transit rail services, while reducing travel time for long distance commuters, also reinforces the concentration of finance, insurance, real estate, and related service industries in central locations. When the scale of downtown concentration is large enough, it can support concentrated entertainment, education, tourism, and numerous other economic activities that thrive on large numbers in close proximity. This concentration of business and other activities, in turn, sustains the appeal of high density central city and close-in suburban neighborhoods.

Table 2.6 Household Savings from Livable Neighborhoods, 1990

	Annual	United States
Linked Transit Trips[b]	1,189	24.7 percent
Passenger Miles [b]	9,575	19 percent
Per Person Vehicles Owned	0.00	Base = 0.74[a]
Vehicle Savings per Household	$3,635	
Aggregate Vehicle Savings[b]	$9,614	
Per Person Vehicle Miles	0.00	
Per Person Vehicle Mile Savings	9,213	9,213[a]
Aggregate VMT Substitution[b]	44,101	

[a]For same poverty status mix.
[b]Millions.

Source: Author's analysis of 1990 Nationwide Personal Transportation Survey.

Proponents of new rapid rail systems often advertise these "land use" benefits. Research on the impacts of the 20-year-old Bay Area Rapid Transit (BART) system in the San Francisco Bay area supports the claim that rapid transit can increase residential density, make neighborhoods more attractive, and promote higher property values.[87]

Long established rapid transit systems in Boston, New York City, Philadelphia, Chicago, and Cleveland may appear superficially to represent substandard support for modern city life. Decades of insufficient recapitalization have left America's most important concentrations of transit service in obvious disrepair. Yet, despite its woeful countenance, the rapid transit system serving New York City, for instance, is still among the local economy's most important assets. Infrastructure improvements to bolster the New York economy capitalize on rapid transit. The success of the project to redevelop the Times Square retail trade area, for example, depends in part on the "200,000 commuters who pass through the Port Authority Bus Terminal each workday and 340,000 daily transit riders who use the times Square subway station of which an estimated 60,000 exit at 42nd Street".[88] The modernization of New York City's rapid transit rolling stock, equipment, and facilities since the 1970s has consistently won the financial support of city and state decisionmakers and bondholders.

Summary

There are several reasonable ways to measure and calculate the benefits of transit to households. This analysis settled on a fairly comprehensive measure: savings from the avoidance of vehicle ownership by households with transit riders. These savings are smallest in households which use transit to bypass congested freeways. They are greatest for households which use transit more generally, and whose above poverty status would ordinary lead to higher auto ownership. The overall household vehicle ownership and operation savings in 1990 were $19.8 billion.

Transit Policy Functions: Private Benefits and the Public Interest

Transit operating and capital costs in 1995 totaled $24.2 billion. Service revenues such as fares, advertising, and lease payments generated $7 billion. The remaining $17.2 billion was paid from taxes on motor fuel, retail sales, property value, and income.[89] The taxes were assessed by State and local governments and by special authorities. The Federal contribution was approximately $4.1 billion. As mentioned in the introduction of this report, this magnitude of public financial support is not new and it is not restricted to a few metropolitan areas. It is nearly universal in urbanized areas with populations greater than 50,000. So, the question is not why *should* so many people support transit, but why *do* they support transit.

In the abstract, transit services may seem remote from the majority of Americans. According to the 1990 census, only one in twenty workers regularly used transit to commute to work.[90] While poor and near-poor people are much more likely than other groups to use transit, only 3.5 percent of their 1990 trips were on transit.[91] Less than 10 percent of United States households are located within 6 blocks of transit service in central cities served by subway and elevated transit systems.[92]

Yet, despite the apparent isolation of transit passengers, transit customers pay, on average, only 44 percent of transit operating costs in fares; taxpayers pay the remaining 56 percent. Taxpayers carry the largest burden in the smallest and ostensibly least important transit systems. Transit serving urbanized areas with less than 200,000 population recovers only about 18 percent of its operating costs from fare revenues.[93] Why are taxpayers so "generous" to transit services?

For each of transit's market niches there are specific constituencies that realize specific indirect benefits that arise from the use of transit by others. These benefits are "divisible" or "club" benefits in that individuals or jurisdictions can "opt in or out" of the social costs and benefits. For example, motorists who customarily are stuck in traffic receive measurable time savings benefits from the ability of rapid transit to help reduce traffic congestion. Thus, in urban areas with severe traffic congestion, highway commuters count themselves among transit's constituents. Parents recognize the ability of transit to reduce the burden on "the family taxi service" at a low cost. Their outcry is the loudest when even skeletal services are cut back.

In addition to these "proximate" constituencies, transit has a more general constituency—the public interest—as a result of diffuse spillover effects that arise mostly from transit's dampening effect on auto ownership and use. Generally, less auto travel means less infrastructure cost, air pollution, fossil fuel consumption, and urban sprawl.

The interests of transit's public constituencies[94] are examined below. Congressional testimony is cited to illustrate the diversity of public goals mustered in behalf of transit funding. The allocation of transit costs and subsidies among transit's three functions is provided as an indicator of their relative priority in local budgetary processes. This is followed by an effort to express in monetary terms one major constituent benefit for each transit function. This is followed by estimated pollution savings, a measurable spillover effect, achieved by each transit function.

The Goals Behind Public Transit Support

George M. Smerk observed in 1976, "In the Congress, transit is considered one of the safe issues; almost nobody is against it".[95] As shown earlier in

Figure 1.1, from 1979 to 1995, State and local financial support for transit operations increased by 214 percent in constant 1992 dollars.[96] In 1995, 467 local transit systems across the United States received financial support for operations from local and State governments. The amount of State and local aid for daily transit operations in 1995 was $9.5 billion.[97]

Congressional testimony through the years exemplify the diverse roles for transit that have resonated with United States leaders and taxpayers.

> " . . we have come to the realization that moving masses of people, not merely vehicles, must be our prime objective. To attain this objective, we realize, belatedly, that adequate, safe, attractive, and convenient mass transit facilities are indispensable, particularly offstreet, grade-separated transit in the areas of heaviest concentration of population. Public aid to transit has been too long neglected".
>
> –Hon. Otto Kerner, Governor, State of Illinois, 1963.[98]
>
> "The ways that people and goods can be moved in these areas will have a major influence on their structure, on the efficiency of their economy, and on the availability of social and cultural opportunities they can offer their citizens. Our national welfare, therefore, requires the provision of good urban transportation, with properly balanced use of private vehicles and modern mass transport, to help shape as well as serve urban growth".
>
> –President John F. Kennedy, 1962.[99]
>
> "I remember years ago when I first came to this program I just made a check on what was the largest city which did not have public transportation, and it turned out to be Odesa, Texas. In other words, we go down a long way before the local politician is ready to give up the service, and I think it is because of the underlying realization that it is a social program".
>
> –Robert McManus, Acting Administrator, FTA, 1993.[100]
>
> "I think there is an asset to public transit that is different than even how much [revenue] you bring in. It is how many cars you take off the road. It is the environmental aspects. It is others. So, just a fare box collection which is important doesn't [by itself] tell the other story".
>
> –Hon. Frank R. Wolf, U.S. House of Representatives, 1996.[101]

> "We believe that there are multiple benefits afforded to us when we invest in transit. Providing an affordable, high quality alternative to the automobile for commuting to work and to other services, transit reduces traffic congestion, improves traffic travel time for motorists, and reduces auto-related air pollution and fuel consumption. Transit also provides low cost mobility for people who cannot afford to own or are unable to drive a car. It improves the vitality and productivity of neighborhoods and business centers".
>
> –Hon. Gordon J. Linton, FTA Administrator, 1996.[102]

Federal, State and local transit legislative documents reflect a similar diversity of goals that motivate public support. Apart from the rhetoric, widespread and sustained State and local transit support suggest that benefits to passengers indirectly serve the interests of larger publics. Transit budgets constitute a linkage of transit's direct beneficiaries with wider publics who perceive an indirect self-interest in providing transit services. Public legislative hearings on transit budgets such as those cited above reveal this interchange between the passenger savings, group specific indirect benefits, and more diffuse benefits of transit services.

Transit's public policy outcomes cluster around the three basic market niches in which transit services are able to attract most of its passengers. That is, transit produces benefits for society at large and for specific nonpassenger constituencies by virtue of its ability to substitute for autos in the three ways described above: low cost mobility, bypassing congestion, supporting transit-oriented neighborhoods. The monetary value of these indirect benefits will be considered below. First, however, it is necessary to consider the relative costs of diverse transit services.

The Costs of Transit Functions

The analysis would be incomplete without consideration of the relative costs that transit incurs on behalf of taxpayers. Transit costs vary according to the internal economics that govern transit service deployment in each market niche. The benefits also vary by market niche. From the foregoing discussion, the financial savings that households secure by substituting transit services for vehicle ownership appear to exceed the fares they pay. In 1990, auto ownership savings associated with low cost regional mobility were $9.3 billion (Table 2.3). Auto ownership savings for the households of commuters circumventing traffic congestion were

$1.1 billion (Table 2.5). The auto avoidance savings for transit users residing in transit-oriented neighborhoods were $9.6 billion (Table 2.6). Considering some overlap among these categories (namely, transit passengers not of working age in above poverty households—low cost mobility—residing in livable neighborhoods), the total household savings in 1990 exceeded the $5.8 billion total transit fare revenues.

Not all transit costs the same. As a consequence of the varying composition of transit capital and operating costs of transit in performing its public policy functions, the net costs to taxpayers (after deducting fare and other operating revenues) are not evenly distributed across transit's functions.

Table 2.7 reproduces the findings of a transit cost study performed in 1992 by Brian McCollom and Lewis Polin. The summary table demonstrates wide disparity in operating cost per passenger according to service "type" across several United States transit systems. Table 2.7 demonstrates that local, radial, and crosstown transit services cost less to

Table 2.7 Types of Transit Services: Operating Cost per Passenger

1990	Types of Transit Routes			
Transit System	Local and Radial	Crosstown	Express	Suburban
Miami	$1.59	$1.92	$9.66	$2.26
Minneapolis	$1.66	$2.40	$3.41	$5.98
Los Angeles	$1.22	–	$2.94	–
St. Louis	$1.34	$1.15	$4.89	$4.82
San Diego	$1.26	$1.89	$2.81	$2.19
Albany, N.Y.	$1.61	$5.11	$1.60	$2.19
San Antonio	$1.01	$1.25	$2.33	$1.90

Source: Federal Transit Administration, *To Classify Transit Services*, (Washington, D.C.: 1992).

operate per passenger than express and suburban services. This reflects the familiar perception that transit services operating in denser downtown neighborhoods are cheaper to operate than sprawling services in low density suburbs. Table 2.7 also demonstrates the extent of disparity among route types even within the same system.

If transit fares were set according to these variations in service costs, the cost disparities would not matter to the taxpayer. But with the important exception of commuter rail services,[103] cost-based (e.g., distance-based) fares are the exception in public transit. As a result, the subsidies for different transit service types are important in the local budget process.

Transit cost disparities are shaped by two basic characteristics of most transit routes.[104] First, transit routes that tend to have high patronage in one direction and low patronage in the reverse direction are said to be "unidirectional". Unidirectional routes incur high net costs as a result of "deadheading" (operating with few or no passengers) in the reverse direction. Deadheading increases the ratio of nonrevenue service to revenue service and thus increases the cost per hour of revenue service.

Secondly, the disparity between peak and off-peak period services increases the ratio of premium pay hours (for split and extended shifts) to regular pay hours for vehicle operators. The so-called "peaking problem" is considered to be transit's Achilles Heel. Casual observers may see crowded transit vehicles during rush hour as a sign of economic vitality. Unfortunately, the situation is often quite the opposite. To supply services to routes with crowded rush hour services, transit managers are forced to increase the disparity between peak and off-peak services. As a result, the new costs tend to exceed the new revenues.

The highest costs per passenger in Table 2.7 reflect the tendency for express bus services to be both highly unidirectional and highly peaked. In six of the seven metropolitan areas reported, express services in 1990 cost more than twice local and radial services on a per passenger basis. Such cost differences are further complicated by each transit mode, e.g., bus, rapid rail, having its own unique internal economies. For example, commuter rail trips cost the most because the fixed capital costs of rail systems are high, there is virtually no off-peak patronage, trip lengths are very long, and deadheading is very high.

In the years before auto ownership was the norm for American households, "off-peak" transit services produced enough surplus revenues to offset the deficits incurred by supplying peak services. The bankruptcy of transit in the 1960s was, in large part, the result of losing off-peak customers to the automobile and to the migration of households from high density central city neighborhoods to suburban neighborhoods.[105]

Speaking of the erosion of transit patronage by mass auto ownership and suburbanization, one may visualize the crisis of private transit companies as a beach on which a retreating ocean has left sharp peaks and valleys along the shoreline. The predominant market niches that remain today are those that have survived the receding ocean by earning the enduring support of local electorates. The three niches that remain represent transit's increasing specialization in patronage and public support.

These three "specializations" have disparate cost patterns, based on differential degrees of peaking, deadheading, and average trip distance. Low cost mobility tends to predominantly a role for bus services. Commuter rail services are pre-eminently used for bypassing highway traffic. Rapid rail services are strongly associated with walkable neighborhoods and commercial centers. The need to use transit to bypass traffic congestion is concentrated in the morning and evening rush hours. Low cost mobility needs tend to be less peaked. In 1985, Charles River Associates developed a model for allocating costs—and subsidies—among income groups according to transit mode, time of day, and trip distance.[106] This model is used below to allocated costs—and subsidies—among transit's three market niches.

Constituencies for Low Cost Mobility

Table 2.8 presents data on the operating and capital costs of low cost mobility transit services. As in Table 2.3 discussed earlier, 2.1 billion annual transit trips were taken in 1990 by people below and above working age and by people near or below poverty with either no car or no drivers license. The 1990 costs allocated to these trips were $4.9 billion, of which $2.2 billion were recovered in fares and other operating revenues, leaving $2.6 billion as the aggregate public expense for low cost mobility.[107]

In 1990, low cost mobility services cost taxpayers about $1.31 per trip. What specific benefits did this low cost mobility confer on nonriders? Chapter 4 of this book includes a calculation of the impact of low cost transit mobility on Federal social service programs such as food stamps and Medicare. These savings result from the transit's ability to reduce the transportation-related costs of Federal program delivery to clients. For example, the Medicare program saves money when patients are able to visit a clinic by public transit rather than being picked up by a clinic van or having a medical house call. It is calculated that each dollar of Federal expenditure on transit services in 1993 produced $0.60 in cost savings for Federal social service programs. If this ratio is extrapolated to all public expenditures on transit services that provide low cost mobility, the savings in Federal social programs alone in 1990 would have been $1.6 billion. Further research is needed to calculate comparable savings for State, local and other public social services in which client mobility is an important cost factor.

Table 2.8 Allocated Transit Costs and Revenues for Low Cost Mobility, 1990

	Annual	Percent of United States
Linked Transit Trips[b]	2,081	43.2
Total Costs[b]	$4,912	37.4
Total Operating Revenues[b]	$2,183	37.9
Total Public Subsidy[b]	$2,729	37.0
Fare Revenue per Trip	$1.05	
Public Subsidy per Trip	$1.31	
Annual Public Sector Budget[a] Savings[b]	$1,638	

[a]Cf., Chapter 4. Estimated effect of one dollar of transit on Federal social budgets: $0.60
[b]Millions

Basic mobility transit services provide an low cost option for people termporily without their cars. It is very difficult to estimate the low cost mobility benefit of transit for the occasional transit passenger. However, research on the responsiveness of transit passengers to different fare levels indicates that the occasional passenger is willing to pay twice or more the normal fare.[108] Many of the 35 million Americans who use transit do so only occasionally. But when they ride transit, they save on taxi fares (average cost: $8.00), car rentals, and the unrecorded cost of inconveniencing their family, friends, neighbors, and colleagues. One needs merely to observe crowding on transit vehicles during snow emergencies and other highway breakdowns to form a subjective judgment on the value of transit as a backup mode for motorists. Transit systems have been enlisted to evacuate hospitals under threat of forest fires, toxic fumes escaping from train wrecks, and floods. Transit systems stood in when earthquakes wrecked major highway arteries in San Francisco and Los Angeles. School budget savings result from the use of transit by children to get to school. Such hard to measure benefits to nonpassengers need to be measured to better focus transportation investments where they are most valued.

Transit as a provider of low cost mobility has become institutionalized in nearly all United States urban areas over 50,000 population and in many smaller urban and rural areas as well. This basic mobility role for transit is implicit in the fact that transit has been a principal target of Federal legislation in 1973 and 1991 to make our society and economy accessible to people with disabilities.

All of this argues that patronage as such is not the principal goal of transit in fulfilling basic mobility. The more important goal has been to connect as many transit dependent people as possible with as many useful destinations in the region as possible.

This would not eliminate patronage as an objective in systems devoted solely to basic mobility. The purpose of getting more riders is to get more revenues and thereby reduce the burden on the taxpayer. Far from begrudging transit riders the modest fare subsidies they receive, many citizens view modest transit fares as an excellent way to reduce the tax burden. In the smallest transit systems, legislators are grateful for the meager 18 percent of costs that are recovered from the farebox.

Reflecting on transit's basic mobility role, the common sense distinction between fare revenue and tax revenue begins to vanish. Transit riders pay both the tax and the fare and, since riders of basic mobility transit services tend to be struggling to make ends meet, taxes and fares both are exactions. In such circumstances, the distinction between subsidy and fare is oftentimes without a difference. To the extent that this distinction is meaningless, transit policy boards are wise to maximize the modest public subsidies available for transit better to maximize transit benefits to their regular clientele.

Constituencies for Congestion Management

The most commonly cited reason for the public's support for transit funding is transit's ability to contend with congested freeways. The "national" transit crisis in the mid-1950s that eventually precipitated Federal financial intervention to rescue the transit industry arose from the desire of commercial railroads to close deficit-prone passenger services that carried commuters to New York City. When the original Federal Transit Act was adopted by Congress in 1964, 40 Republican Congressmen voted against the opposition of their Party leadership to support the program, largely in the name of preserving commuter rail services.[109] The traffic congestion that plagues America's most economically vibrant metropolitan areas continues today to generate support for transit systems designed to combat congestion by offering motorists a bypass option.

Table 2.9 presents data on transit's role in providing a means by which millions of commuters regularly bypass congested freeways to get to work. In 1990 1.5 billion rush-hour transit trips were taken on rail or bus systems by vehicle owners, 57 percent of them working age licensed drivers with household incomes above poverty.[110] These are considered highway "bypass" trips because they are made in the peak by licensed drivers and nondrivers who had potential use of a household automobile. While these highway "bypass" trips accounted for 32.1 percent of overall transit trips, they accounted for 38.7 percent of costs and 41.8 percent of public subsidies in 1990. Congestion management alone cost transit over $5 billion in 1990, of which $2 billion was paid for by passengers, leaving $3 billion be paid by taxpayers. This meant that each congestion bypass trip on transit cost taxpayers $2.00.

Table 2.9 Allocated Transit Costs and Revenues for Congestion Management, 1990

	Annual	Percent of United States
Linked Transit Trips[b]	1,543	32.1
Total Costs[b]	5,088	38.7
Total Operating Revenues[b]	2,008	34.8
Total Public Subsidy[b]	3,080	41.8
Fare Revenue per Trip	$1.30	
Public Subsidy per Trip	$2.00	
Time Savings[a] for Motorists[b]	$7,715	

[a]Motorist travel time savings for each person who bypasses congestion on transit: $5.00 (see footnote 65 in text).
[b]Millions

Time savings for motorists resulting from 1.9 million commuters using transit regularly to bypass congested highways were worth $7.7 billion in 1990. This estimate is based on continuing United States Department of Transportation efforts to estimate the value of time for peak commuters.[111]

Studies have consistently shown that the demand for peak period rapid transit services is comparatively insensitive to fare increases; it is price inelastic.[112] This means that fares are low relative to the value commuters place on these services. In economists' terms, it means that rapid transit passengers receive benefits above the fares they pay. These "consumer surplus" benefits are, at a minimum, the travel time savings the passenger realizes in the equilibration that has occurred between the modes. Thus, transit passengers and motorists share an identity of interests in the travel time (and parking cost) savings that they both realize as a consequence of the subsidized bypass mode.

Constituencies for Livable Communities

People who reside in central cities or in retirement communities, near college campuses, or in other neighborhoods with high quality transit services adapt household economies to the substitution of transit or walking for many errands that are normally achieved by driving. Transit passengers in such "livable communities" saved $9.6 billion (as shown in Table 2.6).

Table 2.10 shows the cost side of the equation. The total cost of transit services for "livable neighborhoods" was $3.14 billion in 1990. Operating revenues and subsidies were both about $1.57 billion. The auto expenses that are avoided in transit oriented neighborhoods and business centers (including $990 million in avoided auto costs of nontransit riders) barely scratch the surface of the advantages enjoyed by households and businesses in these areas. Many of these benefits are captured in the market value premium realized by residential and commercial real estate in transit oriented areas. These real estate premiums are discussed at length in Chapter 5 of this book.

Apart from their economic role in metropolitan areas, high density neighborhoods and business centers are important concentrations of metropolitan political influence. Their political influence flows from a number of assets. Geographic concentration itself is an important asset for assembling numerous common interests arising from proximity and for facilitating communication and coalition formation. Central city public and private sector elites retain consideration influence over regional public finance issues. Regional news media give disproportionate weight to central city issues although their readership is decidedly suburban. Just as the suburbanization of former transit users resulting in transit specialization, city centers have become more specialized in the characteristics of residential and commercial neighborhoods. Sharper ethnic, class, and social lines have been drawn among central city neighborhoods. Increasingly, neighborhoods are able to mobilize their residents politically on behalf of the neighborhood values.

Table 2.10 Allocated Costs and Revenues for Livable Neighborhood Transit Services

	Annual	Percent of United States
Linked Transit Trips[b]	1,189	24.7
Total Costs[b]	$3,142	23.9
Total Operating Revenues[b]	$1,574	27.3
Total Public Subsidy[b]	$1,567	21.2
Fare Revenue per Trip	$1.32	
Public Subsidy per Trip	$1.32	
Vehicle Savings[a] for Neighbors[b]	$990	

[a]499,624 zero vehicle households with no transit trips, central city urbanized areas (> 50,000 pop.), aged 16-64, > 5,000 population density (sq.. mi.) compared to 0.73 per capita vehicle ownership norm.
[b]Millions

Modern transit systems in the larger metropolitan areas of the United States are radially-oriented by design, to make the historical link between suburban residential areas and central city employment. This hub and spoke image mirrors the historical pattern of financial, economic, and political power once unquestionably focused in central cities. With the suburbanization of employment and other aspects of metropolitan life over the last 25 years, the influence of central cities has waned considerably. Still, however, transit has many friends in influential places who continue to reap identifiable economic and social benefits from the human settlement concentrations that only mass transit permits.

Spillover Benefits: The Diffuse Public Interest Outcomes of Transit

The public interest is advanced by transit services to the extent that transit is able to attract a particular stratum of passengers—namely, those who would otherwise drive. Transit is able to attract these passengers insofar as it provides low cost mobility, helps to manage congestion, and makes neighborhoods and cities more pedestrian-friendly. As defined here, public interest benefits are not divisible among individuals or groups in a political jurisdiction. Rather, they are spillover effects from the pursuit of benefits in markets and in local budget processes. The spillover benefits that are identified by representative institutions create in public opinion a degree of receptivity to modest budgetary support for public services such as transit, law enforcement, scientific research, and clean air.

The public interest benefits associated with transit include reduced energy consumption and air pollution associated with lower motor vehicle use than would be necessary in the absence of transit services. They include the budgetary savings from not having to add more highway capacity in congested urban travel corridors. The general taxpayer benefits from the effect of low cost mobility on access to jobs, and thus reduced welfare and unemployment rolls. Every American household benefits from the orders for goods and services that radiate to the general economy from marketplaces concentrated in our largest central cities, a concentration not possible without mass transit services.

Public interest transportation benefits are difficult to measure. The economy is just too complex and infrastructure effects notoriously defy controlled measurement. For example, in the decades required for the diffuse land use impacts of new rapid transit services to mature, the ever thinning chain of evidence linking effects with transit vanishes.[113] Moreover, economic analysis requires discounting long term benefits.

However, a recent comprehensive study of the "total social costs" of motor vehicle use by Mark Delucchi and his associates[114] calculates the "external" costs that motor vehicle use imposes upon the user, her neighborhood, community, and society at-large. The social cost of four air pollutants (PM, VOCs, CO, Nox) is calculated across vehicle types. Delucchi estimated low and high weights for each pollutant per VMT and low and high *costs* per kilogram of each pollutant.[115]

We can use some of DeLucchi's calculations to estimate the savings to society at large from the lessened vehicle ownership resulting from substitution by transit services. Table 2.11 arrays VMT emission savings across transit's three public policy functions.[116] In congestion management, 19 billion VMTs were avoided in 1990. The total in low cost mobility was 40 billion VMTs and in livable neighborhoods 44 billion VMTs. The total auto VMT's replaced by transit services in 1990 were 103 billion. The low estimate of auto related pollutants thus saved in 1990 was 4.7 billion kilograms, conservatively saving $690 million in 1990.

The high estimate of avoided pollutants was nearly 5.5 billion kilograms, potentially saving as much as $12.2 billion in 1990. The large difference between the low ($690 million) and high ($12.2 billion) monetary savings results from uncertainties in the effects of Particulate Matter and from uncertainties in the value of human life and health.

Table 2.11 Emissions Savings by Transit Function, 1990*

Avoided Volumes (Millions)	Congestion Bypass	Low Cost Mobility	Livable Neighborhoods	Totals
VMTs	19,393	40,802	44,101	104,297
		Net Emissions		
Low Estimates				
Kilograms	867	1,833	1,985	4,685
Social Costs	$124	$270	$297	$691
High Estimates				
Kilograms	1,027	2,167	2,347	5,540
Social Costs	$2,272	$4,790	$5,182	$12,244

*Annex A.

Conclusions

Table 2.12 is the balance sheet for household savings credited to substitution of auto ownership by use of and proximity to transit. Driving a privately owned automobile, often alone, is the norm in American society for most travel. One's own car is the clear winner in the mobility sweepstakes.

However, in three urban market niches, transit services offer a tolerable to excellent substitute for auto ownership for millions of Americans. As aresult, where transit serves, auto ownership is significantly lower than the national average of 0.7 vehicles per capita. Among people who depend on mass transit for low cost mobility, vehicle ownership is 0.36 per capita as compared to 0.68 per capita for the same poverty status groups who don't use transit. This substitution of transit for autos saved lower income households $7.11 billion in auto costs in 1990. Among people who used transit to bypass congested urban highways during rush hours, vehicle ownership was 0.66 per capita, just slightly below the general level of 0.73, saving them $1.12 billion in auto expenses. But these households saved $4.63 billion in time and parking costs. Among households whose income and other demographic features would have encouraged auto ownership, but who appeared to substitute transit for autos by choice, vehicle ownership was 0.00, compared to 0.73 for similar poverty status groups, saving them $8.04 billion in auto expenses. The total direct household saving was $24.14 billion in 1990.

Table 2.12 Summary Balance Sheet for Household Savings and the Public Policy Functions of Transit, 1990

	Transit's Public Policy Functions[a]				
Benefits and Costs (Billions)	Congestion Mngmt	Low-cost Mobility	Livable Nghbrhd	Sub Totals	Grand[b] Totals
Transit Passengers	1.54	2.08	1.19	4.81	5.87
Vehicle Ownership	$5.75	$9.29	$9.61	$24.65	$30.08
Fares Paid	($2.18)	($2.18)	($1.57)	($5.94)	($5.94)
Net Benefit	$3.57	$7.11	$8.04	$18.71	$24.14

[a]Calculated from Tables 2.3, 2.5, 2.6, 2.8-2.10, 2.11.
[b]With the restoration of missing cases to the NPTS database.

In providing a substitute for automobile ownership and use, transit wins over identifiable constituencies which realize discrete indirect benefits. These constituencies account for transit's enduring and widespread successes in local budget processes year after year since transit became a local public service one quarter century ago. As shown in Table 2.13, in 1990, local and State constituencies provided $7.38 billion in subsidies to transit. The limited benefits to local and State constituencies calculated in this study totaled $12.62 billion, leaving $5.24 billion in net benefits.

Low cost mobility reduces social and educational expenditures for public and private agencies meeting the specialized needs of low income families, elderly people, students, and other groups. These social budget savings in 1990 are estimated at $1.64 billion. People who use transit to circumvent congested freeways produce daily time savings for motorists on those freeways, apparently $7.71 billion in 1990. Residents of areas heavily served by transit, many of whom do not use transit, are able to accomplish many errands on foot because economic activities are more concentrated in these neighborhoods. People who resided in "livable" neighborhoods but did not use transit saved $0.99 billion in 1990 auto expenses.

Table 2.13 Summary Balance Sheet for Local Jurisdiction "Club" Savings and the Public Policy Functions of Transit, 1990

Benefits and Costs (Billions)	Transit's Public Policy Functions[a]				
	Congestion Mngmt	Low-cost Mobility	Livable Nghbrhd	Sub Totals	Grand[b] Totals
Motorists' Time Savings	$7.71			$7.71	$9.41
Reduced Social Service Costs		$1.64		$1.64	$2.00
POV Savings: Neighbors No Transit Trips			$0.99	$0.99	$1.21
State–Local Subsidies	($3.08)	($2.73)	($1.57)	($7.38)	($7.38)
"Club" Net Benefit	$4.63	($1.09)	($0.58)	$2.97	$5.24

[a]Calculated from Tables 2.3, 2.5, 2.6, 2.8-2.10, 2.11.
[b]With the restoration of missing cases to the NPTS database.

The general public consequences of avoided vehicle miles appear in Table 2.14. Significant benefits accrue to society at-large from the lower vehicle miles traveled by households which substitute transit for automobiles. These benefits include reduced noise, air pollution, energy consumption, highway infrastructure, and other hard-to-measure "external social costs" of auto use. In reduced emissions alone, it appears that transit produced cost savings ranging from $690 million to $12.2 billion in 1990.

From the overall summary presented in Table 2.15, it appears that the valuation of transit depends very heavily on the purpose or policy function transit is being measured against. Just as importantly, transit's valuation depends on whether the beneficiary is the household, the local community, or the larger society. A proper accounting of transit's value, therefore, requires valuation from the perspective of each purpose cross-tabulated against each market niche or policy function. The same, incidently, must be said of other complex activities which governments undertake precisely because benefits and beneficiaries, however obvious, are very difficult to measure, capture through prices, and assess in the marketplace.

If this analysis is correct, transit's net total benefits in 1990 to households, local communities, and the nation as a whole totaled at least $26 billion and perhaps as much as $40 billion.

Table 2.14 Summary Balance Sheet for Spillover Public Interest Savings and the Public Policy Functions of Transit, 1990

	Transit's Public Policy Functions[a]				
Benefits and Costs (Billions)	Congestion Mngmt	Low-cost Mobility	Livable Nghbrhd	Sub Totals	Grand[b] Totals
Harmful Emissions Avoided					
Low Estimate	$0.12	$0.27	$0.30	$0.69	$0.84
High Estimate	$2.27	$4.79	$5.18	$12.24	$14.94
Federal Subsidies	($1.70)	($1.50)	($0.80)	($4.00)	($4.00)
Low Estimate	($1.58)	($1.23)	($0.50)	($3.31)	($3.16)
High Estimate	$0.57	$3.29	$4.38	$8.24	$10.94

[a]Calculated from Tables 2.3, 2.5, 2.6, 2.8-2.10, 2.12.
[b]With the restoration of missing cases to the NPTS database.

Further Research

Low cost mobility is transit's pre-eminent "social" function. Congestion management is transit's "transportation system management" function. Supporting transit oriented neighborhoods and commercial centers is transit's "urban development" function. These three functions pervade the literature and legislation on transit. Yet, too little has been done to measure transit's functions for the public. With this examination of transit's substitution effect in distinct market niches, we have suggested one approach to this measurement process.

Along similar lines, the succeeding three chapters represent formal economic analysis of transit's benefits to customers and society. Such detailed analysis is necessarily piecemeal. However, these pages instructively respond to the most important substantive and methodological questions about transit benefit measurement.

Table 2.15 Summary Balance Sheet for Net Total Benefit and the Public Policy Functions of Transit, 1990

	Transit's Public Policy Functions[a]				
Benefits and Costs (Billions)	Congestion Mngmt	Low-cost Mobility	Livable Nghbrhd	Sub Totals	Grand[b] Totals
Passenger Household Savings	$3.57	$7.11	$8.04	$18.71	$24.14
Local Jurisdiction "Club" Savings	$4.63	($1.09)	($0.58)	$2.97	$5.24
	Spillover Public Good Savings				
Low Estimate	($1.58)	($1.23)	($0.50)	($3.31)	($3.16)
High Estimate	$0.57	$3.29	$4.38	$8.24	$10.94
Total Benefits					
Low Estimate	$6.62	$4.78	$6.96	$18.37	$26.22
High Estimate	$8.77	$9.30	$11.84	$29.92	$40.32

[a]Calculated from Tables 2.3, 2.5, 2.6, 2.8-2.10, 2.11.
[b]With the restoration of missing cases to the NPTS database.

Annex 2.1 Net Emission Savings from Transit Services

Table 2.16 Low Estimate of Pollution Savings by Transit Functions, 1990*

Avoided Volumes (Millions)		Congestion bypass	Low Cost Mobility	Livable Nghbrhoods	Totals
g/mi	VMT'S	19,393	40,802	44,101	104,297
0.20	PM	3,879	8,160	8,820	20,859
3.10	VOCs	60,119	126,488	136,715	323,321
38.20	CO	740,822	1,558,655	1,684,676	3,984,153
3.60	NOx	69,816	146,889	158,765	375,470
Emissions "Reduced"		874,636	1,840,192	1,988,976	4,703,804
Transit Emissions		7,140	7,598	3,997	18,735
Net Fewer Emissions		867,496	1,832,593	1,984,979	4,685,069
$/kg	VMT'S	19,393	40,802	44,101	104,297
$9.75	PM	$38	$80	$86	$203
$0.10	VOCs	$6	$13	$14	$32
$0.01	CO	$7	$16	$17	$40
$1.17	NOx	$82	$172	$186	$439
Emission Costs Avoided		$133	$280	$302	$715
Transit Emission Costs		$9	$10	$5	$24
Net Emission Savings		$124	$270	$297	$691

*VMTs from Table 2.6, 2.6-2.7 above applied to: Mark A. Delucchi, The Annualized Social Cost of Motor Vehicle Use In the United States , 1990-91: Summary of Theory, Data, Methods, and Results, (Davis, CA: Institute of Transportation Studies, 1996), Report Number 1 in the Series, Table on Page 51. Cost estimates vary with uncertainties in pollution estimates and health impacts.

The emissions per transit passenger mile are from American Public Transit Association, 1993 Transit Fact Book, (Washington, D.C., 1993), p. 21. Estimates are based on typical work trips with national average vehicle occupancy rate. The passenger mile data are from Tables 2.3 to 2.5, resulting in an NPTS total transit passenger mile estimate (50 b) that is 31 percent higher than reported by transit systems.

Table 2.17 High Estimate of Pollution Savings by Transit Function, 1990*

Avoided Volumes (Millions)		Congestion bypass	Low Cost Mobility	Livable Nghbrhoods	Totals
g/mi	VMT'S	19,393	40,802	44,101	104,297
0.30	PM	5,818	12,241	13,230	31,289
3.70	VOCs	71,755	150,969	163,175	385,900
45.30	CO	878,515	1,848,352	1,997,796	4,724,663
4.00	NOx	77,573	163,210	176,406	417,189
Emissions Saved		1,033,661	2,174,772	2,350,608	5,559,041
Transit Emissions		7,140	7,598	3,997	18,735
Net Emissions		1,026,521	2,167,174	2,346,611	5,540,306
$/kg	VMT'S	19,393	40,802	44,101	104,297
$133.78	PM	$778	$1,638	$1,770	$4,186
$1.15	VOCs	$83	$174	$188	$444
$0.09	CO	$79	$166	$180	$425
$17.29	NOx	$1,341	$2,822	$3,050	$7,213
Em. Costs Saved		$2,281	$4,799	$5,187	$12,268
Transit Emis Cost		$9	$10	$5	$24
Net Emission Savings		$2,272	$4,790	$5,182	$12,244

*VMTs from Table 2.6, 2.6-2.7 above applied to: Mark A. Delucchi, The Annualized Social Cost of Motor Vehicle Use In the United States , 1990-91: Summary of Theory, Data, Methods, and Results, (Davis, CA: Institute of Transportation Studies, 1996), Report Number 1 in the Series, Table on Page 51. Cost estimates vary with uncertainties in pollution estimates and health impacts.

The emissions per transit passenger mile are from American Public Transit Association, 1993 Transit Fact Book, (Washington, D.C., 1993), p. 21. Estimates are based on typical work trips with national average vehicle occupancy rate. The passenger mile data are from Tables 2.3 to 2.5, resulting in an NPTS total transit passenger mile estimate (50 b) that is 31 percent higher than reported by transit systems.

Annex 2.2 Household Classification for Transit Benefits

Key Variables

Nationwide Personal Transportation Survey data for 1990 transit passengers were subdivided by five criteria as shown in Tables 2.18; 2.19; and 2.20:

Status of Household: (1) Above Poverty (2) Near or Below Poverty

Age of Traveler: (1) 16 to 64 (2) Under 16 or Greater than 64

Passenger was a Driver: (1) Licensed Driver (2) No Driver's License

Household Vehicle Ownership: (1) No Vehicle (2) One or More

Trip Time: (1) Peak (2) Off-peak

The marker variables for *congestion management* were vehicle ownership and the use of transit during peak periods. The other important variable was above poverty status, indicating that auto ownership was not a strain.

The marker variables for *low cost mobility* were income near or below poverty and no household vehicle. An important exception was made for off-peak trips by people with licenses and vehicles. These were interpreted as "back-up" transit trips, included in the concept of affordable mobility.

The marker variables for *livable neighborhood* transit were above poverty income and no vehicle ownership. Included also were groups that couldn't easily fit into the other two categories.

Classification Issues

Ideally, the classification of transit benefits and beneficiaries would combine household features such as poverty status, personal features such as trip purpose, and transit features such as services on separate rights of way or service headways. When such combinations are possible, a clear delineation of transit's public policy functions should be obvious. In fact, the Federal Transit Administration (FTA) and American Public Transit Association (APTA) are currently working with a number of transit agencies to generate data that would make such combinations possible. A brief analysis of the strengths and weaknesses in the current approach could

Table 2.18 Households Selected for Low Cost Mobility

Poverty Status	Age	Driver	Own Vehicle?	Peak	Households	Transit Trips
			Subtotal		2,306,045	1,262,278,439
Above	<16>64	License	Yes	No	44,950	21,858,549
Above	16-64	License	Yes	No	706,872	312,843,693
Near/Below	16-64	License	Yes	No	64,429	38,535,826
Above	<16>64	No Lic.	Yes	No	440,871	181,676,733
Above	16-64	No Lic.	Yes	No	201,345	101,865,705
Near/Below	<16>64	No Lic.	Yes	No	76,347	41,625,129
Near/Below	<16>64	No Lic.	Yes	Peak	88,626	60,275,341
Near/Below	16-64	No Lic.	No	No	123,154	98,271,744
Near/Below	<16>64	No Lic.	No	No	129,947	85,988,183
Near/Below	16-64	No Lic.	Yes	Peak	83,647	78,084,793
Near/Below	16-64	No Lic.	No	Peak	102,925	69,762,606
Near/Below	16-64	License	No	No	62,410	58,462,389
Near/Below	<16>64	No Lic.	No	Peak	89,426	55,108,254
Near/Below	16-64	License	No	Peak	28,456	26,403,812
Near/Below	<16>64	License	No	Peak	12,554	5,037,266
Near/Below	<16>64	License	No	No	11,895	5,037,266
Near/Below	<16>64	License	Yes	No	10,964	4,307,055
Near/Below	<16>64	License	Yes	Peak	10,964	4,307,055
Near/Below	16-64	No Lic.	Yes	No	16,263	12,827,040

Highlights

Percent below poverty 51

Percent above poverty, licensed, auto owners, offpeak 25

Trips per household 547

Trips per above poverty household 444

suggest data generation strategies for the FTA-APTA enterprise and for others interested in this topic.

An ideal-type classification of *low cost mobility* transit use would focus on the tripmaker's interest in a relatively low fare as such, with little concern for travel time savings. So, the typical low cost mobility passenger pays a low transit fare and gets relatively infrequent service, often with a circuitous route to his or her destination. This service is typically found in transit services serving urban areas with less than 200,000 population.

In Table 2.18, transit trips by people without either a drivers license or a vehicle in their household and "near or below" poverty incomes were classified as low cost mobility trips. The income consideration suggests that vehicle ownership would be a financial strain. Also classified here were trips by people not in the 16-64 "working age" group. Many of these people have above poverty incomes, suggesting some choice in vehicle ownership and acquiring drivers licenses. To the extent that choice is involved, many of these trips could be classified as "livable neighborhood" trips Many children and elderly people have very modest need for motorized travel, so that minimal transit makes the neighborhood quite "livable" for them. Added, too, were off-peak trips by licensed vehicle owners, considered to be cases in which transit is serving as a "back-up" mode for the private vehicle. On closer examination, many of these "back-up" trips occurred in the "shoulders" of the morning and evening peak and thus could be classified equally well as "congestion bypass" trips.

The ideal *congestion management* class would be trips for which the generalized marginal costs were equal between driving and taking transit. Generalized costs are the sum of transit fares, travel time pegged to income, parking fees, and vehicle operating costs. This ideal obtains in discrete travel corridors in a number of urbanized areas. This ideal seldom applies to transit services that do not travel on separate rights-of-way and thus literally bypass congested highways.

In Table 2.19, peak period transit trips by people with one or more household vehicles are counted as bypass trips. The rationale for including bus trips is that rush hour buses, especially in large urban areas, arguably dampen highway travel demand—and thus arguably contribute to reduced congestion—depending on how many intermittent stops the buses add to the traffic flow. Bus trips that offer only inferior travel times, insufficient seating, and other drawbacks could just as well be classified as low cost mobility trips. By contrast, high frequency peak period buses or trains used for very short hops and connections should be classified as "livable neighborhood" services, because the traveler may be getting service superior to private auto travel but is not bypassing congestion per se.

The ideal *livable neighborhood* trip is one made by members of households which, by choice rather than income or other disadvantage, substitute transit and walking for the "norm" of auto dependency. These households reside in high density transit oriented neighborhoods in which many daily errands are achieved on foot rather than by car. For them auto ownership simply offers no set of advantages that equals its annual cost. For them, a car is simply an unnecessary burden—walking and transit an advantage.

Table 2.19 Households Selected for Congestion Management

Poverty Status	Age	Driver	Own Vehicle	Peak	Households	Transit Trips
			Subtotal		1,096,396	889,087,556
Near/Below	16-64	License	Yes	Peak	71,211	48,443,838
Above	16-64	License	Yes	Peak	624,086	520,408,828
Above	<16>64	License	Yes	Peak	13,284	25,101,477
Above	<16>64	No Lic.	Yes	Peak	241,248	178,163,214
Above	16-64	No Lic.	Yes	Peak	146,567	116,970,199

Highlights

Percent trips taken during peak period 100

Percent Yes vehicles in the household 100

Percent of households Above 94

Percent licensed, working age 58

Trips per household 811

In Table 2.20, trips made by above poverty households without vehicles are the essential livable neighborhood trips. Included are trips made by people above and below the 16-to-64 working age cohort. To the extent that these individuals would settle for less dense transit services or lower density neighborhoods, their trips should be classified as low cost mobility trips. Also, many peak trips in this category bypass congested roads and thereby reduce congestion.

The classification of transit trips among the three public policy functions in this chapter is not arbitrary, but neither is it ideal. This approach offers three distinct advantages. First, without ignoring obvious cases in which a given transit trip may actually serve two or three public policy functions, the double counting of benefits is studiously avoided while household (vehicle ownership) and diffuse (VMT-emissions) benefits are all-inclusive. Secondly, the approach is relatively parsimonious, reducing the classification to trip (peak-off-peak), person (licensed driver, age) and household (poverty status, vehicle ownership) characteristics that have obvious connections to transit's ability to substitute for cars. Third, since overlaps are not counted, certain social benefits are undercounted, so that the reported constituency benefit estimates err conservatively.

Table 2.20 Households Selected as Livable Neighborhood Transit Users

Poverty Status	Age	Driver	Own Vehicle	Peak	Households	Transit Trips
			Subtotal		1,515,111	680,968,553
Above	<16>64	License	No	No	17,781	10,852,992
Above	<16>64	License	No	Peak	27,515	11,294,831
Above	<16>64	No Lic.	No	No	184,466	76,746,183
Above	<16>64	No Lic.	No	Peak	96,696	40,872,938
Above	16-64	License	No	No	245,670	111,225,244
Above	16-64	License	No	Peak	377,273	166,988,400
Above	16-64	No Lic.	No	No	309,112	148,095,249
Above	16-64	No Lic.	No	Peak	256,598	114,892,716

Highlights

Percent Above Poverty 100

Percent No Vehicle Households 100

Percent Working Age 79

Trips per Household 449

Notes

51 1990 Nationwide Personal Transportation Survey.

52 Ibid., p. 33.

53 Average Annual Costs of Household Vehicle Ownership, calculated by the author from United States Bureau of Labor Statistics data reported in American Automobile Manufacturers Assoc., *Motor Vehicle Facts and Figures,* (Detroit: AAMA,1995), p. 61.

54 In 1994, the auto industry spent $7 billion informing the public of these virtues.

55 When public works are undertaken to counter-act unemployment, the jobs created by transfer through taxes or borrowing should be considered benefits only to the extent that income transfers are explicit aims of the policy.

56 Op. cit., p. 416.

57 Terminology used by Mancur Olson, *The Logic of Collective Action*, For the economist, the term "constituents" could be replaced with "club members" to recall the economic theory of clubs. Cf., James M. Buchanan, "An Economic Theory of Clubs", *Economica*, (February 1965), pp. 1-13.

58 Altshuler, op. cit., p. 44.

59 Earlier studies have attempted to capture "interest group" benefits in terms of socio-economic characteristics, ignoring the organization of groups in relation to the decisionmaking process. See Robert Cooter and Gregory Topakian, "Political Economy of a Public Corporation: Pricing Objectives of BART", *Journal of Public Economics*, Vol. 13, No. 3 (June 1980), pp. 301- 302.

60 When transit proponents look to Europe for models of effective transit, they find that the densities required for more effective transit--i.e., livable neighborhood densities--are more common in western Europe.

61 This analysis is limited to households with 5 or fewer members because the number of observations recorded for larger households in the Nationwide Personal Transportation Survey base is too small to rely upon.

62 A recent study projected that in the absence of transit in the Philadelphia region auto ownership would be 0.3 vehicles higher per household. The current estimate is more modest. The Urban Institute and Cambridge Systematics, *The Economic Impacts of SEPTA on the Regional and State Economy*, (Washington, D.C.: United States Department of Transportation, 1991), p. 3.9. See also, Meyer and Gomez-Ibanez, op. cit., p. 97.

63 Toll facilities on which High Occupancy Vehicles are not charged, like State Route 91 in Orange County, California.

64 $2,714 annual cost of auto ownership and operation (BLS); 1.74 average private vehicle occupancy in 1992 (BTS); 11,000 annual miles per private motor vehicle (BTS).

65 According to the 1993 American Housing Survey, of 16.6 million households that relocated in the previous year, 514,000 or 3 percent chose their new neighborhood because it was convenient to public transportation. Apart from the desire for good schools, transit access was singled out more often than other public services combined. United States Census Bureau, 1993 AHS-N Data Chart, Why Move? - Table 2.11.

66 Annex B.

67 See Annex B: Classification of Transit Users.

68 Customarily, most accounts of transit trips report vehicle boardings and entries through transit station turnstiles. They are known as "unlinked trips". Survey data, however, tend to report "linked trips", which include more than one transit vehicle boarding.

69 For the likelihood that a household will achieve the normal (0.73) rate of auto ownership, poverty status is a simpler and more powerful measure than household income. Households represented on each of the three transit policy "functions", comprise a difference mix of (1) "above poverty" and (2) "near and below poverty". In all the tables in this chapter, poverty status (so defined) is controlled for in comparing vehicle per person and annual VMTs per person.

70 Operating revenues for these trips in 1990 were $2.2 Billion (Table 2.9).

71 In urban areas with 50,000 to 500,000 population, 1.2 percent of all trips. John Pucher and Fred Williams, "Socioeconomic

Characteristics of Urban Travelers: Evidence from the 1990-91 NPTS", *Transportation Quarterly*, Vol. 46, No. 4, (October 1992), Table II.

72 Cf. M. Millar, Argonne National Laboratory, cited in Jim Burnley, *The Status of the Nation's Local Mass Transportation: Condition and Performance,* Secretary of Transportation's Report to the Congress, (Washington, D.C.: U.S. Department of Transportation, 1988), p. 58.

73 John F. Kain, "The Spacial Mismatch Hypothesis: Three Decades Later", *Housing Policy Debate*, Volume 3, Issue 2 (Washington, D.C.: Fannie Mae, Office of Housing Policy Research, 1992).

74 Altshuler, et al, Op. Cit., pp. 435 - 436.

75 Federal Transit Administration, *Unsticking Traffic: When Transit Works, and Why*, (Washington, D.C.: United States Department of Transportation, 1994).

76 Martin J.H. Mogridge, op. cit.

77 Unfortunately, because the Nationwide Personal Transportation Survey database does not include work locations, Table 2.4 does not permit the isolation of travel to central city jobs, where rail transit is most competitive. However, case study evidence presented in Chapter 4 of this book lends strong support to the view that transit rail commuters experience travel times similar to those of motorists plying the same corridors. That is, the 29 percent difference in speed reported in Table 2.4 is largely attributed to the large proportion of automobile commutes outside the highly congested central city-oriented channels in which transit rail services are most concentrated.

78 See Annex.

79 Operating revenues for Congestion Management in 1990 were $2.2 billion (Table 2.10).

80 American Public Transit Association, *1994-1995 Transit Fact Book*, (Washington, D.C.: APTA, 1995), Table 45.

81 Reported in Chapter 2 below.

82 "Unstick .ig Traffic", op. cit. A reverse dynamic equilibrium can be generated by new highway construction, resulting in slower trips across commuting modes, a phenomenon known as the "Downs-Thompson Paradox".

83 Urban Mass Transportation Administration, Demographic Change and Recent Work Trip Travel Trends, 1985, cited in United States Department of Transportation, The Status of the Nation's Local Mass Transportation: Performance and Condition, (Washington, D.C.: Report to Congress, June 1988), Figure 3.1.

84 John Holtzclaw, "Using Residential Patterns and Transit to Decrease Auto Dependence and Costs", Natural Resources Defense Council, June 1994.

85 Operating revenue for livable neighborhood services in 1990 were $1.6 billion (Table 2.11).

86 Thus, the households savings resulting from reduced vehicle ownership reappear in higher housing costs.

87 John Landis and Robert Cervero, "BART at 20: Property Value and Rent Impacts", prepared for the 74th Annual Meeting of the Transportation Research Board, (Washington, D.C.: January, 1995).

88 42nd Street Development Project, Inc., *42nd Street Now! Executive Summary*, (New York, 1993), p. 11.

89 FTA, 1995 National Transit Database, Tables 1 and 5.

90 Federal Highway Administration, *New Perspectives in Commuting*, (Washington, D.C.: United States Department of Transportation, 1992).

91 Author's analysis of Nationwide Personal Transportation Survey.

92 Ibid.

93 Ibid.

94 Transit's "private constituencies who are suppliers, transit employees, etc. receive financial returns for their sales and labor which, economically speaking, offset each other, so that there is no net "benefit" to them per se. Taxpayer financed jobs, that is, are conventionally considered jobs that otherwise would have been created by private investors elsewhere in the economy.

95 George M. Smerk, "Productivity and Mass Transit Management", *Urban Transportation Efficiency*, (New York: ASCE, 1976), p. 197.

96 American Public Transit Association, *1993 Transit Fact Book*, (Washington, D.C.: APTA, 1993), Table 20.

97 Ibid.

98 United States House of Representatives, Hearings, *Committee on Banking and Currency Hearings on HR 3881*, (Washington, D.C.: GPO, 1963), p. 248.

99 Cited by Robert C. Weaver, Housing and Home Finance Administrator, Ibid., p. 16.

100 United States House of Representatives, Hearings, Committee on Appropriations, on Transportation and Related Agencies Appropriations for 1964, (Washington, D.C.: GPO, 1963), p. 852.

101 United States House of Representatives, Hearings, Committee on Appropriations, on Department of Transportation and Related Agencies Appropriations for 1997, (Washington, D.C.: GPO, 1996), p. 1336.

102 Ibid., p. 1037.

103 Discussed below.

104 John T. Meyer, John F. Kain, and Martin Wohl, *The Urban Transportation Problem*, op. cit., pp. 173 - 181.

105 Richard L. Oram, "Peak Period Supplements: The Contemporary Economics of Urban Bus Transport in the U.K. and United States", *Progress in Planning*, Vol. 12, Part 2, (Oxford: Pergamon Press, 1979), p. 95.

106 Charles River Associates, Inc. (CRA), Allocation of Federal Transit Operating Subsidies to Riders by Income Group, (Washington, D.C.: Federal Transit Administration, Office of Policy Development, 1986).

107 The CRA calculation were on 1983 data. In Tables 2.9, 2.10 and 2.11, 1990 NPTS patronage data and 1990 aggregate cost data were applied to unit cost estimates developed with 1983 data.

108 Richard Oram, *Deep Discount Fares: Building Transit Productivity with Innovative Pricing*, (Washington, D.C.: United States Department of Transportation, 1988).

109 "Congress Clears $375 Million Mass Transportation Bill", *Congressional Quarterly*, (Washington, D.C.: Week Ending July 3, 1964), p. 1340.

110 Bus riders were licensed drivers.

111 Cf., Hon. Rodney E. Slater, *Report on Funding Levels and Allocations of Funds for Transit Major Capital Investments*, (Washington, D.C.: United States DOT, 1997), p. A-9. The calculated value of congestion relief for 1990 is based on 80 percent of the average metropolitan wage rate--$11.70 for 1996.

112 C.f., Charles River Associates, *Public Transportation Fare Policy*, (Washington, D.C.: United States Department of Transportation, 1977), Chapter 3.

113 Cf., Cambridge Systematics, "A Review of Methodologies for Assessing the Land Use and Economic Impacts of Transit on Urban Areas", FTA Discussion Paper (Washington, D.C., FTA, 1995).

114 Mark A. Delucchi, *The Annualized Social Cost of Motor-Vehicle Use In The United States, 1990-1991: Summary of Theory, Data, Methods, and Results*, (Davis, CA: Institute of Transportation Studies, 1996), Report Number 1 in the Series, Table 1.10.

115 Delucchi attributes the lion's share of difference in his estimates to unknowns in the effect of certain pollutants and uncertainties in the monetary valuation of human life.

116 Using VMT's avoided by the use of transit from Table 2.3, Table 2.5 and Table 2.6 above.

3 Public Transit for Congestion Management

"Or even—how's this?—they drive over to Lex and Fifty-ninth, by private car or cab, and take the downtown IRT subway. There's a local stop on Spring Street, less than two blocks from Maitland's studio. By taking the subway, they eliminate the risk of getting stuck in traffic. And I think they could make the round trip in ninety minutes to two hours, allowing five or ten minutes for killing Maitland."

"I don't know," Delaney said doubtfully. "It's cutting it thin".

"Want me to time it, sir"? Boone asked, getting a little excited about his idea. "I'll drive from Dukker's place to Maitland's studio and back, and them I'll try the same trip by subway. And time both trips".

"Good idea," Delaney nodded. "Make both between ten and three on Friday, when the traffic and subway schedule will be approximately the same as they were then."[117]

Introduction

This chapter is presented in two parts. The first part, "The Problem of Auto Traffic Congestion," examines how the urban congestion problem is addressed in traditional transportation planning and policy making. In the United States transit is seen as a means of marginally diminishing roadway traffic in congested corridors. The traditional view holds, however, that transit has relatively little influence on the underlying factors that generate the congestion problem. Most analysts regard transit as a palliative rather than a real solution. The second part of the chapter, "The Role of Modern Transit in Managing the Congestion Problem", challenges the traditional view. There we show how existing rapid transit systems exercise measurable control over congestion in the United States, and do so cost-effectively. This chapter demonstrates how transit investment and targeted fare and subsidy policies can measurably control traffic congestion probably for generations to come.

The Problem of Auto Traffic Congestion

Congested conditions on roadways represent the rationing of transportation system capacity. In the absence of road pricing, consumers use the transportation system in excess of the efficient level of demand resulting in inefficient levels of congestion. Stiglitz (1986) describes roadways as publicly provided private goods. "There is a large marginal cost associated with supplying additional individuals" the specified good, and "if a private good is publicly provided, there is likely to be over consumption of the good".[118] Button (1993) argues "when users of a particular facility begin to interfere with other users because the capacity of the infrastructure is limited, then congestion externalities arise".[119] Arnott and Small (1994) assert that "an externality is brought about when a person does not face the true social cost of an action".[120]

Congested roadways are evidence of over-consumption of the publicly provided transportation network. Every additional individual imposes a cost on pre-existing road users greater than the cost he or she perceives. Drivers are not faced with the 'true' costs of their consumption. They experience no price restrictions with regards to their consumption, hence roadways are over consumed resulting in congestion.

Moore and Thorsnes (1994) argue because "some people cannot be excluded from using or consuming some public goods, the government has difficulty using price as a signal to guide production decisions".[121] In other words, the lack of road pricing in addition to contributing to congestion, creates a barrier in supplying or creating roadway capacity. Without price signals, real demand is difficult to calculate. When demand is unknown, it is almost impossible to estimate equilibrium supply, hence appropriate roadway capacity is also unknown.

Congestion Trends

Constructing more highway capacity has reached a point of greatly diminished returns in developed metropolitan areas. Not only are cities running out of space for new lanes, but the continued addition of highway capacity may be paving the way to a larger, less-treatable gridlock.[122]

Downs (1992) asserts that to eliminate congestion would involve changing deeply imbedded physical and social characteristics and structures of American metropolitan areas as well as Americans themselves. It would require "...persuade[ing] millions of Americans to alter some of their most cherished social goals and comfortable personal conduct".[123] Table 3.1 illustrates the desire to travel is growing rapidly, enhanced by population growth.

Table 3.1 Trends in Congestion in the 10 Largest Metropolitan Areas

Metropolitan Area*	Congestion Ranking in 1990	Travel Growth 1982-1990	Pop. Growth 1982-1990	Daily VMT per Lane-Mi 1982-1990
		(Percent Change)		
New York	9	31.3	0.7	16.5
Los Angeles	1	46.2	15.4	27.2
Chicago	5	49.4	6.1	25.9
Philadelphia	17	48.1	3.7	22.6
Detroit	14	12.1	5.0	-2.4
San Francisco	3	47.5	10.5	35.8
Washington	2	57.5	27.0	28.1
Boston	16	35.8	3.9	26.1
Houston	10	33.9	19.5	-4.1
San Diego	6	83.6	29.2	61.8

*In order of 1990 population.

Source: David Lewis, "Curbing Gridlock: Peak-Period Fees to Relieve Traffic Congestion", National Research Council, Vol. 2. 1994, p. 107.

Downs (1992) formulated four short-term causes of congestion existing and reinforcing each other: rapid population and job growth; more intensive use of vehicles; failure to build new roads and the lack of market pricing of scarce capacity. The amount of congestion in an urban area is intuitively related to its population. Larger urban centers tend to be more congested and typically have a range of solutions to address transportation problems, indicating a recognition of the problems of relying on roadway solutions.[124]

Downs (1992) found that population in eighteen metropolitan areas increased between 1980-90 by 45 percent or more. In 1983, interstate mileage was approximately 43,000 miles and in 1990 was approximately 45,000 miles, a growth of about 4 percent.[125] Between 1983 and 1990 vehicles per household grew by 5.14 percent. The number of persons and households grew at rates of 4.34 percent and 9.34 percent respectively.

The Nationwide Personal Transportation Survey (NPTS) in Table 3.2 also shows that employment growth is more than 3 times population growth. Even with stationary population growth, the number of cars increased. Studies have suggested that auto ownership per household is 1.77 in 1990.

Table 3.2 Household Vehicle Ownership, 1983 - 1990

	1983		1990		Total	Urban
(Thousands)	Total	Urban	Total	Urban	Change	Change
Households	85,371	55,857	93,347	58,977	9.34	5.59
Persons	229,453	146,180	239,416	138,910	4.34	-4.97
Lic. Drivers	147,015	92,574	163,025	100,827	10.89	8.92
Workers	103,244	66,541	118,343	76,397	14.62	14.81
HH Vehicles	143,714	87,011	165,221	98,675	14.97	13.41
Veh. per HH	1.68	1.56	1.77	1.67	5.14	7.41

Source: *1990 Nationwide Personal Transportation Survey*, Report No. FHWA-PL-94-018, June 1994, Table 2, p. 9.

Intensity of Vehicle Use The second cause of growing congestion is more intensive use of vehicles. "The fraction of all households owning 2 or more vehicles rose form 29 percent in 1969 to 53 percent in 1988".[126] Many scholars[127] have concluded that the privacy, comfort and convenience of vehicle usage makes its very difficult for other modes to compete. Downs (1992) suggested that vehicle usage was also affected by the growth of suburbs ill served by public transit, the dispersal of job locations and commuters switching to autos from public transit as their preferred commuting mode.

Lagging Highway Capacity Failure to build more roads allows congestion to increase. The Texas Transportation Institute (TTI) in analyzing 50 cities–their respective Roadway Congestion Index (RCI), Vehicle Miles Traveled (VMT) Growth, and roadway capacity growth–found that total driving increased more than twice as fast as roadway capacity.[128]

Table 3.3 is an excerpt of the RCI for ten metropolitan areas taken from a TTI analysis of congestion in 50 metropolitan areas. These ten urban areas were randomly chosen to illustrate congestion growth. An RCI of 1 or greater represents an urban area with undesirable congestion. The TTI analysis indicates 56 percent of the urban areas studied, experienced levels of undesirable congestable (RCI of 1 or greater) and an additional 32 percent experienced RCI levels of .9 or greater.

Table 3.3 Congestion Index for Selected Urban Areas

Urban Area	1982	1983	1984	1985	1986	1987	1988	1989	1990	Change
Pittsburgh	0.78	0.76	0.76	0.78	0.79	0.79	0.81	0.82	0.82	5%
New York	1.01	1.02	0.99	1.00	1.06	1.06	1.10	1.12	1.14	13%
Boston	0.90	0.93	0.95	0.98	1.04	1.04	1.12	1.09	1.06	18%
Baltimore	0.84	0.84	0.85	0.84	0.88	0.90	0.92	0.99	1.01	20%
Denver	0.85	0.88	0.93	0.96	0.97	0.95	0.99	1.01	1.03	21%
Columbus	0.68	0.71	0.71	0.71	0.75	0.78	0.79	0.82	0.83	22%
Chicago	1.02	1.02	1.05	1.08	1.15	1.15	1.18	1.21	1.25	23%
Dallas	0.84	0.89	0.94	0.98	1.04	1.02	1.02	1.02	1.05	25%
Los Angeles	1.22	1.27	1.32	1.36	1.42	1.47	1.52	1.54	1.55	27%
Washington	1.07	1.09	1.12	1.20	1.28	1.30	1.32	1.36	1.37	28%

Source: Texas Transportation Institute Analysis-excerpt *from Estimates of Urban Roadway Congestion-1990,* p. 18.

Researchers doubt that building more highways can significantly reduce congestion. Mogridge[129] and research contained in this report suggest that, in certain cases, only transit investments can improve travel times. For the past several years, researchers of traffic systems have observed that in congested urban corridors served by a dedicated guideway transit mode, door-to-door journey times tend to be equal. The findings have profound implications for transportation investment strategies in congested urban corridors. The results favor a transit-led strategy of investment for the improvement of system performance by all modes. The data suggests that capacity growth is waning behind the growth in vehicle, worker and licensed driver numbers.

However, increasing capacity as a sole response to alleviate congestion is not a feasible alternative. To illustrate this point, "... New York would require 201 additional lane-miles of freeway and 257 lane-miles of principal arterial streets per year to maintain the 1990 congestion level with the 3.4 percent growth in DVMT [daily vehicle miles traveled] it experienced between 1987 and 1990".[130]

Chronic Traffic Congestion as a Pricing Problem The underlying cause of severe traffic congestion, apart from the sizable influence of unpredictable traffic accidents and breakdowns, is the absence of marginal cost pricing for times when road space is most scarce due to heavy travel demand. In market economies, valued services and goods are apportioned by prices which tend to reflect different valuations. Services whose costs, and commercial value, are affected by seasonal or time of day peaks in demand, such as hotels, restaurants, movie theatres, and airlines, generally sharply increase their prices during the peak periods. In this way, equilibrium between suuply and demand is maintained, and the higher costs of meeting peak demand are covered by receipts.

Queuing of customers is a classic symptom of market disequilibrium. When traffic on multiple highways, each with several lanes in one direction, is routinely reduced to a snail's pace, all the motorists in the corridor are paying the same price for their commute—in the form of delays. This is the condition of most congested highways in industrial societies. While highly egalitarian—in some sense—this condition is highly inefficient.

If a driver is willing to pay a fee to be on time for an important event with his or her kids, or to a particularly important meeting at the office, the option is not available except on the High Occupancy Toll (HOT) lanes on State Route 91 in Southern California. Since the motorist pays the same price regardless of differences in the value of a bridge at different times of the day, there is every incentive to cross the bridge at the most valuable times. The most valuable time, of course, is during the busiest times of the day for the bridge, when its space is the scarcest. In addition, most people driving to work receive free parking, which tends to be worth the most in the most congested cities, and free parking also adds value to the bridge one crosses to reach the parking space.

Some have argued over when motorists pay the full costs of the transportation facilities they use. This argument is beside the point. Since nearly every adult in the U.S. owns a car and drives every day, the full costs of highway use are passed on to motorists in poor air quality, in travel delays, in highway construction, in the car damage caused by poorly maintained highways, in neighborhood blight, and in taxes. There is no choice in the matter. It is only a technicality to say that motorists do not pay the full costs of mispriced highways. The more telling issue is a consequence of the lack of choice.

Without marginal cost pricing, motorists have no opportunity to save money by avoiding heavy traffic. The are denied the option to purchase access to a fast lane when it matters most, and to otherwise express preferences and achieve economies with one's buying power. All trip purposes are treated equally. Moroever, since any motorist shares the costs no matter when he or she drives, the motorist will tend to drive when to do so is most valuable to him or her. Obviously, because of work and other inflexible schedules, the most valuable times to travel coincide for many drivers. Traffic jams are the inevitable result.

Unable to break the grip of this daily transportation vise, commuters have turned to markets where their earning power can make a difference. The choice of where to live presents commuters with opportunities to use their financial resources to find good schools for their children, improve the quality of the neighborhood, and create new transportation options. These choices, ironically, contribute to traffic congestion in another way.

Long-Term Causes of Congestion A major long run cause of congestion is the increasing spatial separation of work and residences. Workers tend to live further away from Central Business Districts (CBD) and places of work, seeking private, quite and spacious single detached homes. Workplaces are also locating away from the CBD where more ground space is available and at lower costs. These aspects reduce the feasibility of commuting by mass transit and discourage the use of car pooling and other ride sharing. Although future locations may be influenced or regulated by future urban planning and development, existing locations cannot relocate easily. "To persuade more commuters to shift modes without changing the locations of their homes or jobs, it would be necessary to make net benefits of solo driving less than those of travel by other modes".[131]

Congestion Management Alternatives

If motorists were to pay the incremental costs of their specific driving choices, it is argued, they would economize in their daily driving habits, \there would less traffic and the road system will be less congested.[132] Until government agencies remove the obstacles to road pricing, however, alternative measures are in order. In travel corridors with the most severe traffic congestion, the construction of new roads or lanes is ineffective because latent travel demand fills them up until the pre-existing amount of travel delay (i.e., congestion cost) is restored. The only way to reduce congestion ...is to introduce better public transport facilities which reduce the number of people who travel by car on the roads.[133]

Meyer and Gomez-Ibanez (1981) analyzed methods designed to manage and/or alleviate congestion. Their results are arrayed in Table 3.4. The authors emphasize faults experienced under application of these alternatives are due to wrongful implementation of policy makers and are not inherent. "Projects may concentrate on local areas, ignoring impacts throughout the traffic network; thus some projects may have been frustrated or offset by others".[134]

Johnson (1994)[135] also examined alternatives to road-pricing that manage congestion. He considered enhancing public transit, designing intermodal systems facilitating movement between modes and encouraging transit-oriented land development. Transit could be enhanced by increasing its speed within urban core areas, making it the preferred mode in high-density areas. Instituting priority lanes for public transit within an urban core would increase transit speeds.

Movement between modes may be facilitated by intermodal systems that provide 'park-and-ride' lots at rail stations and bus access. Essentially a rider has access to three modes of travel; drive to the station, ride a train to a partial destination and board a bus to the final destination.

Transit-oriented land development is not extensive in the United States. Generally land development in the United States is characterized by "..low density settlement patterns and isolation of residential areas from shopping, services and jobs; poor public transit service; and pedestrian and bicycle-unfriendly residential and shopping areas".[136]

However, in some European countries and cities, land development is quite transit oriented. In Amsterdam for example, residents have voted for a 'car-free' zone, which prohibits vehicles from traveling at speeds in excess of 18 mph and has increased parking fees. "The city will expand rail-lines and provide car parks near terminals on the periphery of the core"[137] to encourage non-automobile travel.

Table 3.4 Alternatives to Road Pricing

Method	Advantages	Disadvantages
Traffic Control	Augment highway capacity by better timing of traffic signals. Coordinate lights to reduce queues, using electronic sensors. Costs relative to gains are modest.	Operating costs of monitoring, repairing and adjusting the management system during its life cycle.
Reduce Street Parking	Reduce parking on congested streets—may add a lane or more of traffic in each direction.	Costs of replacing on-street parking with parking lots. Lost retail sales from loss of customer on-street parking.
One-Way Streets	One Way street direction eliminates left turns against traffic, reduces accidents and provides for greater signal optimization.	O-D distance offsets speed increases. Transit must differ in arriving and leaving directions. Reduced business access.
Reverse Rush Hour Lanes	Increase of 20 percent and higher of traffic volumes in peak direction.	Small speed reductions in other direction. Depends on imbalanced traffic flow[138] Road space must by pass bottlenecks and allow two lanes in other direction.
Meter Traffic	Control the ramp access with gate and lights. Increases speeds. In Los Angeles, speeds increased from 15-20m/hour to 50 m/hour.[139] Ideal for supersaturation.	Delays at ramps and alternative streets offset some time savings. Need capacity at ramps and alternative streets for delayed and diverted traffic.

Downs (1992) suggests that schemes designed to remedy congestion should be formed to alleviate rather than eliminate congestion. Alternatives should: reduce the duration of the period of maximum congestion; increase the average commuting speed; increase the proportion of all commuters traveling during peak hours; and, reduce the intensity of commuter frustration.[140]

Strategic Corridors and Congestion Management We assume that people travel by the best method available to them. They will therefore prefer to travel by individual vehicles until the journey attractiveness becomes as bad as that by shared vehicles. The journey speeds using shared vehicles therefore set the journey speeds using individual vehicles if there is not enough space for everyone to travel by individual vehicles.[141]

A strategic corridor is defined as an urban corridor with severe congestion and served by rapid rail. In such a corridor the time it takes to complete a journey, door to door, tends to be the same across different modes of transportation. Smeed and Wardrop (1964),[142] Sharp (1967),[143] Goodwin (1969)[144] and Webster and Oldfield (1972)[145] all agree that overall journey speeds can be increased by limiting road capacity and switching displaced drivers to mass transit. Kain (1994) also explains that "higher levels of transit demand in a corridor provide numerous opportunities for cost savings and improvements in trip speeds".[146]

Downs (1992)[147] suggested that an equilibrium point between the costs of public transport and vehicle usage exists resulting in the presence of travelers that are indifferent between these modes of travel. Modal speed becomes the determining factor of usage.

Button (1993) observes that the nominal cost of each mode may actually form a small proportion of the factors influencing modal choice. Mogridge (1990) analyzed survey data observing mode choice behavior and discovered evidence that travelers do switch between modes. Tables 3.5 through 3.8 present evidence of mode switching behavior. "The proportion switching is 81 or 84 in 494 (16.4 or 17.0 percent) in the first pair of surveys and 58 or 60 in 399 (17.1 or 17.7 percent) in the second pair. One can only assume that a much larger proportion than 15 percent of journeys are potential mode-switching journeys".[148] Evidence that mode switchers do exist help to explain the "phenomenon" that it is mass transit which determines the "critical speed" of travel. Suchorzewski (1973)[149] states that the critical speed–the lowest acceptable road speed–is dependent upon the speed and efficiency of public transport, that it is public transport that has to be improved if traffic speeds are to be increased.

Triple Convergence Downs (1992) performed extensive congestion studies which lead to his discovery of the theory of triple convergence. Drivers search for the shortest, quickest, least congestion route and in so doing, converge together creating the exact congestion they are trying to avoid. When congestion occurs, some drivers switch to alternative routes and modes, freeing up space on the congested route. The new space increases

Table 3.5 Numbers and Percentages of Respondents by the Different Kinds of Mode-Choice Behavior Over a Week-Long Period for the Four Panel Data Sets

	Autumn 82	Spring 83	Autumn 83	Autumn 84
Uni-modal*	906 (86.2%)	1,177 (86.2%)	929 (86.2%)	1078 (84.9%)
Modal-mix	157 (14.8%)	204 (14.8%)	161 (14.8%)	192 (15.1%)
Total	1,063	1,381	1,090	1,270

*The uni-modal category includes respondents who traveled by a single mode over seven days.

Source: Mogridge, op. cit., p. 210.

Table 3.6 Mode-Choice by Full- and Part-Time Workers

Proportions[a]	Full-time	Part-time
Uni-modal[b]	85.4%	84.5%
Modal-mix	14.6%	15.5%
Total	787 (100%)	122 (100%)

[a]Total part- and full-time workers 909, total sample size 1090 respondents in autumn 1983.
[b]The uni-modal category includes respondents who traveled by a single mode over seven days.

Source: Mogridge, op. cit., p. 210.

speeds and attraction to other drivers. Downs (1992) explains that with the expansion of congested roadways--designed to relieve congestion and increase speeds--three types of convergence occur: Spatial, Time and Modal Convergence.[150]

Spatial Convergence occurs when "many drivers who formerly used alternative routes during peak hours switch to the improved expressway".[151] This leads to less congestion and increased speeds on alternative routes, however, expressway congestion increases and expressway speeds decrease. The second type of convergence occurs when drivers that formerly traveled before or after peak-periods switch to driving during rush hours. This convergence is designated *Time Convergence*.

Table 3.7 Percentage of Respondents by the Various Kinds of Mode-Mix Behavior

	Autumn 1982	Autumn 1983
Bus-car	17.2%	17.4%
Bus-rail	7.0%	15.5%
Bus—other[a]	10.8%	12.6%
Car-other	19.8%	14.1%
Car-rail	22.9%	22.4%
Rail-other	10.8%	9.3%
Other[b]	11.5%	8.7%

[a]Other mode, includes walk, cycling etc.
[b]It includes people who were observed to have a choice-set of more than two modes.

Source: Mogridge, op. cit., p. 210.

The third and final convergence is termed *Modal Convergence* and denotes commuters that switch from mass transit to driving during peak-hours because roadway travel has become faster. Convergence of this type leads--as do the others--to increased congestion and slower speeds on the "improved" expressway.

Although the only remedy Downs (1992) foresees that alleviates congestion while avoiding triple convergence is tolling the roadways, the network may be improved without tolls by increasing overall speeds through mass transit improvements.

The triple convergence theory lays the groundwork necessary to observe that modal choices are interconnected; relative speeds are often the influencing factor of choice. Thomson (1977) states that "..the quality of peak-hour travel by car tends to equal that of public transport...all efforts to improve peak-hour travel by car will fail unless public transport is also improved".[152] He reiterates what others like Suchorzewski (1973) state regarding the Downs--Thomson Paradox that increasing roadway capacity leads to transit patrons switching to vehicles, leaving transit fares to increase and service to diminish. The result is a new equilibrium which leaves both modes worse off then previously experienced.

Table 3.8 Modal-Choice Behavior, Comparisons Over Time

Number of Respondents	Spring 1983			Autumn 1983		
	Modal -mix	Uni-modal*	Total A82	Modal-mix	Uni-modal*	Total A82
Modal-mix	24	57	81	23	35	58
Uni-modal*	60	353	413	37	244	281
Total S83	84	410	494			
Total A83				60	279	339

*The uni-modal category includes respondents who traveled by a single mode over seven days.

Source: Mogridge, op. cit., p. 210.

If congestion cannot be eliminated, rather alleviated, and increasing speeds in the entire network does alleviate congestion, then policy should be created to do just that. Accepting that it is mass transit that determines network speeds, naturally leads to a policy that improves the speed of mass transit thus increasing speeds throughout the entire network.

Second Best Policy Response If optimum congestion tolls were charged, the motorist might well choose to use mass transit, to join an auto pool, to make the trip at a less congested time, or to use a less congested but perhaps more circuitous route.[153]

Pigou (1920)[154] advanced the idea of pricing resources such that marginal costs are equal to average costs. Goods exhibiting increasing costs need to be taxed and goods exhibiting decreasing costs subsidized, to ensure that marginal and average costs are in equilibrium. Thus to avoid over consumption (congestion) of roadways, a tax should be levied on users. These taxes could take the form of increased gas or vehicle registration prices, or tolls. When tolling is infeasible, subsidizing transit becomes a second-best solution.

Button (1993) states that "...in a situation where marginal cost pricing is not universal and where political expedience leans against the introduction of measures such as road pricing, subsidies may offer a pragmatic second-best approach to the externality problem".[155]

Public transit in general experiences scale economies. Scale economies occur when costs per unit diminish if the scale of operations increase; i.e. the greater the number of passengers the lower the per passenger cost of operation. Edgeworth (1925)[156] and Hotelling (1932)[157] explained that by levying the optimum tax on one mode results in decreasing the costs of both modes, "...a subsidy paid to public transport with a downward sloping cost curve, would lower the costs of travel on both".[158] In the reverse, a policy that increases roadway capacity luring away transit patrons, increases the costs of transit with increasing per passenger costs of the road system.

Moore and Thorsnes (1994) argue that funds have been over allocated to highways to fix congested roadways. They suggest that mass transit subsidies would decrease the gap drivers perceive between the costs of driving and the costs of transit, by making transit faster, more reliable and more convenient. Rail, the authors feel, eventually will take on a substantially larger ridership for managing congestion. "Rail has the potential to offer greater convenience and amenity to riders than the bus".[159] Road users are subsidized, paying too low a price relative to the social costs of using the roadway during rush hours, resulting in too many cars on the roads and too few transit riders, "giving much slower speeds than could be achieved if prices were set at resource costs".[160] To reduce private vehicle use, Buchanan (1963)[161] recommended good cheap public transport coupled with explanation of the subsidies. Table 3.9 shows examples of estimated subsidies provided to drivers and transit patrons.

Eliminate Congestion? Congestion cannot be eliminated but can be managed by implementing the appropriate policy. In strategic corridors it is accepted in the prevailing literature that transit is the speed domineering mode and that to improve network speeds, policy should be designed to improve the speed of mass transit. In urban areas experiencing severe congestion, and road pricing is not an option, second best theory indicates that subsidizing transit is the second best method to manage congestion.

"In the long-run, the most potent factor in maintaining a 'ceiling' on private car traffic in busy areas is likely to be the provision of good, cheap public transport, coupled with the public's understanding of the position".[162]

Table 3.9 Implicit Subsidies Compared

	Driver Pays	Driver Subsidies	Transit Fares	Transit Subsidies
Individual's Direct Cost	[a] $0.36–$0.46/mi. (@ 15,000 mi./yr.)[163]		About $1.00 per trip[164]	
Government's Direct cost (Capital & Operating)	$44.3 billion, or 60% of costs[165]	$29 billion, or 40% of costs (WRI)	$6.3 billion, or 30% of costs	$15 billion, or 70% of costs
Society's Indirect Costs due to air pollution, congestion, parking and other factors		[b] $310-$592 billion, or 27-37% of total annual auto systems costs[166]		[b] $2.1-$5.3 billion, or 6-'8% of total annual transit system costs
Total Subsidy Estimates (NRDC)		[c] $378-$660 billion annually		[c] $19-$22.2 billion annually

[a] Price ranges are for a 1995 Ford Escort LX, 1995 Ford Taurus GL, and a 1995 Chevrolet Caprice Classic.
[b] Total system cost estimates are: $1.2-$1.6 trillion for auto; and $27.8-$30.9 billion for transit.
[c] NRDC totals include a $72 billion government automobile subsidy estimate and a $16.9 billion government transit subsidy estimate. The transit subsidy includes a pro rata share of road and highway expenses.

Source: "Transportation System Subsidies Compared-MTDB in San Diego Quantifies Subsidies", *The Urban Transportation Monitor,* Vol. 9, No.20, (Oct. 27, 1995), pp. 1, 4.

The Role of Modern Transit in Managing Auto Traffic Congestion

This book introduces two approaches to understanding the dynamics of urban congestion that give new importance to the role of transit in managing the problem. Based on the concept of dynamic equilibrium, the first approach is based on important new findings about the ability of high capacity transit systems (such as subways and light rail systems) to actually regulate and equalize the degree of congestion in all surface modes in a multi-modal corridor. The second approach draws on the economic concept of "the second best". This approach takes a new look at the common but disputed argument that transit subsidies, by lowering fares, encourage transfer from private vehicles, alleviating the congestion externality. The findings show that present levels of subsidy in the United States can be fully justified on the basis of their congestion-reducing effects, despite low cross-elasticities.

Dynamic Equilibrium in Traffic Congestion

For the past several years, researchers of traffic systems have observed that in congested urban corridors served by a high-capacity transit mode, door-to-door journey times tend to be equal. New research postulates an economic theory and empirical evidence supporting these observations. The findings have profound implications for transportation investment strategies in congested urban corridors and favor a transit-led strategy of investment for the improvement of system performance by all modes.

In general, the amount of time it takes to make a trip during peak hours, and the number of users who decide to use roads versus transit, depend on a number of factors: the highway capacity, the costs of using a car versus taking public transit, and individual traveler's tastes. In spite of all of these variables, a travel pattern emerges in congested urban corridors: the time it takes to complete a journey, door-to-door, tends to be the same across different modes of transportation. Furthermore, it is the journey time by the transit mode that seems to determine the journey time for other modes. In fact, this pattern of converging travel times is predicted by economic theory.

Current planning practice usually does not allow for the convergence of travel times and, in fact, proceeds quite differently. The standard planning practice consists first of predicting the number of trips that will be made between two locations, based on the number of inhabitants in both places, the location of jobs, etc. Then, these trips are apportioned among the different modes based on the traveler's income, personal tastes, etc. It is at this point that standard practice departs from the theoretical and empirical results set forth in this chapter: The standard approach does not account for travelers who move back and forth between modes, much as motorists move between lanes on a highway in their search for a faster-moving lane. It is the presence of these "explorers" that allows for the travel times to converge across modes, toward those for transit.

The omission of inter-modal effects leads to the following result: benefits to users of other modes brought about by the improvement to the peak-hour performance of transit (by increasing investment, say) are neglected. One possible solution is to apportion total trips among modes such that the travel times are the same across all modes.

Benefits of Transit Rail Investments Not Captured in Benefit-Cost Analysis
The current practice of benefit-cost analysis as applied to transit investments follows a conventional planning approach. Total demand is forecast as trips between zones. These trips are then allocated to modes using another model. Typically, the benefits from the proposed transit investment are estimated as the willingness-to-pay[167] for the trips taken, plus the benefits of reduced congestion on the highways. Recent studies conducted for the Federal Transit Administration's Office of Policy[168] have identified three areas in which this model fails to capture the full array of benefits from transit investment.[169]

First, there remains the issue regarding the interaction between transportation investment and land use. The planning approach described above was used to justify numerous road projects by assuming, for instance, that an outlying area would be developed. Under this assumption, build and no-build scenarios were compared and road projects were shown to display strong benefits. Of course, development in the outlying area would likely never occur without the road project. The no build scenario, in effect, assumed development that would not occur without building the road.

Furthermore, the conventional approach does not adequately address the issues of whether the planned road actually contributed to net new development, or, whether the development was preferable to other development alternatives. In contrast to highways, the benefit-cost of transit rail investments does not account for the transit-oriented development which would legitimately be associated with a "build" scenario. A refinement of methods is underway which incorporates: interactive land use and transit development scenarios; (hedonic) methods for assigning values to development alternatives; and yet other (stated preference) methods which seek to indirectly gauge the benefits of transit-oriented development.

The second area of benefits not captured by existing transit benefit-cost analysis is cross-sectoral resource savings. The absence of transit restricts the mobility of some,users and may require an increase in resource use for medical and social services. Studies demonstrating these benefits have been conducted in the U.K. and methods for incorporating them into transit benefit-cost analysis are being developed.

Finally, conventional transit benefit-cost analysis does not account for the inter- (or multi) modal interrelationships which are seen to exist in congested urban corridors. Mogridge[170] (1991) has shown that in congested urban corridors, door-to-door journey times are nearly equal and tend to converge to the journey time of the transit mode. New evidence confirming this finding has been documented in recent and ongoing studies in the United States (see Table 3.10).

Triple Convergence or Travel Time Convergence? Downs (1992)[171] discusses as a principle of traffic analysis the notion of "triple convergence" whereby peak hour traffic speeds converge spatially (across the road network), in time and across modes. Under the triple convergence principle, an improvement in peak-hour travel conditions on high-capacity roadways "...will immediately elicit a triple convergence response, which will soon restore congestion during peak periods, although those periods may now be shorter". The prospects for improving transportation performance through transit investment are just as promising. Downs states that a new fixed-rail public transit system should initially reduce peak-period traffic congestion, "...[b]ut as soon as drivers realize that expressways now permit faster travel, many will converge...onto those expressways during peak periods".

However, in congested urban corridors the observed convergence of peak-hour, door-to-door journey times—by the highway and transit modes—suggests that a different dynamic is at work. If the travel time convergence dynamic were in effect, a carefully chosen fixed-rail investment would indeed yield an improvement in journey times by highway. In general, the convergence of all journey times to the journey time by the transit mode implies that a change in the performance of transit will result in a change in the performance of highways.

This phenomenon of travel time convergence to the transit journey time has profound policy implications for the planning and allocation of funds for transportation in metropolitan areas. Furthermore, it enables the application of benefit-cost methods to alternatives across different modes, i.e., highway and transit projects are more readily comparable insofar as the cross-modal impacts can be compared where the conditions for trip time convergence are found to exist.

Modal Explorers What explains the phenomenon of travel time convergence? One claim is that a dynamic relationship exists which parallels that of a multi-lane highway. Speeds across lanes tend to be equal because some drivers are "explorers" who seek out the faster-moving lane thus driving the system to an equilibrium speed shared by all lanes. By the same token, in congested urban corridors some travelers and commuters are explorers who value travel time improvements highly. They are not committed through circumstance or strong preference to either mode and they behave as occasional mode switchers.

If the transit mode has a high-speed, non-stop segment, then the door-to-door journey time by this mode will be relatively stable and small shifts in ridership will not significantly impact the journey time by the transit mode. On the other hand, under congested conditions even a one-half percent increase in highway traffic volume in the peak period can have a major impact on journey times.

Because the journey time by transit is stable and determined by the speed of the high-capacity mode, transit "paces" the performance of the urban transportation system in the congested corridor. The modal explorers, like exploring drivers on the multi-lane highway, serve to bring about an equilibrium speed across modes as they seek travel time advantages across modes.

Travel Time Equilibrium and Modal Choice While travel time represents a dominant component in the cost of trips, the generally accepted models of modal choice and the assignment of trips to networks would not predict travel times to be equal. Rather, the theory behind current practice is that individuals choose a mode based on income, car ownership, price differentials and modal preferences which account for non-money factors like convenience, uninterrupted travel, etc. The persistence of equal, or near equal, travel times across modes in congested corridors suggests that current theory fails to correctly capture modal interrelationships in a multi-modal system.

Annex 3.1 presents the economic theory for consumer behavior under congestion and develops the conditions under which door-to-door trip time by highway converges to the trip time by the high-capacity transit mode. It further demonstrates how congestion promotes the modal explorer behavior.

Empirical Evidence Economic theory tells us that if congestion is severe enough, then journey times will tend to equal the journey time by the transit mode under the assumption of growing marginal disutility. This assumption can be tested empirically by estimating the relationships between travel time differentials, congestion and additional factors.

Source of Data In an ongoing study for the Federal Transit Administration, door-to-door travel time tests were conducted on 17 urban corridors. The testing was conducted between February and October of 1995. The corridors were selected based on criteria which included: congestion, population density, the existence of mature dedicated-guideway transit systems, and public transportation headways. A list of the seventeen corridors where data was collected is given in Table 3.10. The corridors span a range of moderate to high congestion. In each corridor random routes of origins and destinations were selected. Survey crews conducted peak hour trips on the different modes under comparable conditions. Over 1000 trips were recorded and some of the average results are reported in Table 3.11. Of the trips taken, 570 pairs of comparable auto/transit trips were observed. Congestion data for the metropolitan areas in which each of the corridors was taken from the recent TRB study on urban congestion.[172] The Metropolitan Planning Organizations in each corridor provided information on transit headways.

Analysis of Data A regression analysis of time differentials was conducted, presented in Table 3.12. The absolute value of the travel time difference, auto vs. transit, was regressed against the metropolitan area congestion index and the transit mode headway (minutes).

As explanatory factors, congestion and headway do little to explain the variation between each of the 570 trip pairs. This is not surprising since these variables have no variation within the corridor and transit mode. However, we observe that the coefficient for congestion is negative while that of headway is positive and both coefficients are significant at the 99 percent level. This means that as congestion increases and as transit headways decrease, the travel times between the automobile mode and the transit mode become more equal.

Table 3.10 Strategic Transit Corridors Measured

Corridor	Modes Measured
Atlanta--I-20	Auto, Heavy Rail, HOV
Atlanta--I-85	Auto, Heavy Rail
Boston--Mass Pike	Auto, Commuter Rail
Boston--Southeast Expressway	Auto, Heavy Rail
Chicago--Midway	Auto, Heavy Rail
Chicago--O'Hare	Auto, Heavy & Commuter Rail
Cleveland--Brook Park	Auto, Heavy Rail
Philadelphia Schuylkill--Bryn Mawr	Auto, Commuter Rail
Phila. Schuylkill--Upper Merion	Auto, Commuter Rail
Philadelphia--Wilmington	Auto, Commuter Rail
Pittsburgh--Parkway East	Auto, Express Bus
Princeton--New York	Auto, Commuter Rail
San Francisco--Bay Bridge	Auto, Commuter Rail
San Francisco--Geary	Auto, Express Bus
Washington--I-66	Auto, Heavy Rail, HOV
Washington--I-270	Auto, Heavy Rail
Washington--I-95 Woodbridge	Auto, Commuter Rail, HOV

There undoubtedly are additional factors which contribute to the explanation of travel time differentials, some of which are location-specific while others are associated with price and other variables. However, wefind that the evidence supports the theory that in congested urban corridors the growing marginal disutility from time spent traveling causes door-to- door journey times to converge to the journey time by the high-capacity, transit mode. Furthermore, the data show that reducing transit headways contributes to shorter trip times and also contributes to a reduction in the time differentials between modes.

The analysis above tells us that the observation of equal or near equal travel times across modes is consistent with consumer theory and may be observed under a wide range of circumstances with high levels of congestion. Congestion, if severe enough, will drive a multi-modal transportation system towards convergent travel times. Further empirical study of congested corridors will reveal which combination of underlying factors (economic, demographic, spatial-locational, etc.) are most closely associated with the condition of travel time convergence. Travel time convergence in congested urban corridors and the factors promoting that convergence should be crucial elements in the development of transportation policies. This is especially true in an environment of budgetary restraint and limited congestion pricing.

Table 3.11 Door-To-Door Travel Times for Peak Journeys

Corridor	Auto Mode (Minutes)	Transit Mode* (Minutes)
New York, Jamaica, Queens- Midtown Manhattan	63.9	64.4
San Francisco Bay Bridge	72.3	73.1
Phila. Schuylkill Expressway--Bryn Mawr	48.4	52.5
Chicago--Midway	54.2	60.6
Chicago--O'Hare	53.9	59.3
Pittsburgh Parkway East	38.1	42.5
Princeton--New York City	113.4	104.9
Washington--I-270	71.9	67.4

*The transit mode is assumed to be a "high-speed", fixed guideway mode. This mode can include dedicated high occupancy vehicle (HOV) lanes.

Source: "Unsticking Traffic"[173] and ongoing research.

Implications for the Benefit-Cost Analysis of Transit Projects The benefit-cost analysis of transit investment examines the demand for trips and derives consumer surplus estimates based upon the schedule of demand. Non-transit trips are mostly assigned to the highway network and cost savings from reduced congestion are estimated. Trips are allocated between modes using a modal choice algorithm which does not take into account the door-to-door dynamic equilibrium between the modes. When the allocated trips are assigned to the highway network, this means that even under highly congested conditions, forecast journey times for road and transit are likely to be highly divergent.

Table 3.12 Strategic Corridor Regression Results

Dependent Variable: Absolute Value of Trip Time Difference

(Auto–Transit)

Variable	Coefficient (t-values)
Constant	15.30 (5.54)
Congestion Index	-3.48 (-2.45)
Headway	0.506 (7.80)

All coefficients are significant at the one percent level

Summary Statistics

Number of Observations	570
R^2	0.098
Mean Dependent Variable	15.68
F-Statistic	30.97

As a first step towards refining the benefit-cost analysis of transit investment with a view to accounting for the phenomenon of convergence in congested corridors, the analyst should examine whether apportioning of travelers among modes leads to convergent journey times once the travelers (or trips) have been assigned to the urban transportation network. If the corridor under analysis is one in which convergence is likely to occur then there is strong theoretical and empirical justification for calibrating the modal constants in the modal choice model such that the assignment of traffic yields near equal journey times.

Second Best Policy Response

Transit and highway together comprise a system of urban transportation. The policy imperative for transportation is to recognize the reality of political and institutional barriers and to achieve efficient use of society's resources subject to those constraints. The lack of road-pricing is a significant constraint on the efficient use of the transportation system. The result is over-use of the road network and inefficient levels of congestion. In the absence of road pricing, the optimal policy response is to subsidize the transit mode.

Prospective commuters weigh the benefits and costs that they face when choosing between alternative modes (i.e., transit or highway). The costs which are not fully borne by the individual, including the congesting effect of an additional private vehicle on the road, do not generally enter his/her calculations. "Second-best" policy options account for the reality that the congesting effect of travelers do not enter private calculations in the absence of road pricing.

Economic theory then suggests that all travelers, whether car users or not, can be made better off[174] if the new users are made to pay some special toll, for instance, that covers the additional social costs they impose. Of course, such a "congestion" toll (also known as road pricing) is unlikely, owing to institutional and political barriers. In effect, there remains a distortion in the price of road travel created by uncompensated social costs. Failure to address this price distortion leads to inefficient levels of congestion and slower travel times. Attendant negative effects include time and productivity losses for road users, higher costs of production, and higher levels of pollution.

In such a world, all travelers can still be made better off (if not as well off as in the "first-best" world) if prices in other, related sectors of the economy are similarly distorted.[175] Indeed, the solution may seem paradoxical: if road travel is underpriced (i.e., there exists no road pricing to discourage additional congestion), then it is justified to underprice the cost of travel on other modes. More precisely, fares on public transit would be subsidized, so that transit users would pay less than what it costs to transport them.

The reasoning behind this seeming paradox is the following: a subsidy draws potential auto travelers to transit, thus averting additional congestion.[176] In fact, in the absence of road pricing, subsidizing some travelers not to use roads makes everyone better off, road and transit users alike as long as the subsidy is less than the congestion costs imposed by each additional driver. Subsidies are set such that, for the last prospective auto traveler they attract to transit, they exactly offset the additional congestion costs to all current road users that would have occurred had that person decided to use a car instead.

The next section reviews research on the pricing of public transit. It focuses on how both road and transit users' welfare can be maximized in the absence of road pricing.

Background Studies by Glaister,[177] Lewis[178] and others in Europe during the 1970s and 1980s indicate that subsidizing transit services is an economically efficient response to the political and institutional barriers to road pricing. The authors adapted the general theory of "second best" in a methodology for calculating the optimal subsidy for public transit when roads are systematically under-priced. This methodology combines cross-price elasticities[179] of the various transportation modes with the social marginal costs of each into a model that generates the optimal set of prices for each mode.

The argument in favor of subsidizing public transit follows from the under-pricing of road travel. In the absence of marginal-cost pricing, individual drivers do not take into account the congestion (or social marginal cost) they impose on others when making travel decisions. The theory of second best says that, when prices deviate from their marginal (social) cost in one sector, then using marginal-cost pricing in other, related sectors will not lead to a social optimum. In the case at hand, road use is under-priced due to lack of tolls and intense congestion levels. Indeed, society would be better off by subsidizing transit fares, thus drawing travelers away from road use and reducing the social costs they previously imposed on other road users.

Opponents of subsidies to public transit argue that cross-price elasticities are so small and transit's share of the total transportation market so insignificant that most analysts assumed the optimal transit subsidies derived by this approach would also be insignificant.

New evidence from Dr. Mohring and Dr. David Anderson[180] suggests that, even in lightly congested urban areas such as Minneapolis/St. Paul, the optimal congestion toll on highways may be as high as $0.49/mile or more. Estimates of this magnitude suggest that the social costs of driving may be higher than previously thought. If the cross-price elasticities between road use and public transit are significant, then the subsidization of public transit would lead to large (social) cost savings.

Furthermore, research by Hickling Lewis Brod Inc. indicates that a certain class of travelers may be especially sensitive to relative prices among transportation modes. These travelers are "explorers" who frequently switch modes. They are sensitive to price when choosing travel modes and might switch to transit based on the optimal transit fare.

Taken together, the high social costs of congestion, the presence of "explorers" and the historically significant cross-price elasticities between road use and public transit argue for a thorough examination of transit subsidy policy.

Efficient Subsidies Currently, transit operations and capital spending are subsidized by many levels of government. Since the distribution of transit operating funds does not currently account for the economic efficiency of the "second best" optimal subsidy, it is likely that some of these current subsidies are too low or too high to yield the economically efficient fare structure given road underpricing.

Subsidies can only be "efficient" in an economy operating in what economists have called a "second best" world. The "first best" policy response to underpriced of roads would be to price roads at the social marginal costs of driving. Marginal cost pricing is a fundamental indicator of the efficient allocation of resources. The theory of "second best" suggests other outcomes when this "first best" outcome is unachievable.

The theory of "second best" states, generally, that when a distortion (underpriced roads) exist in one sector, traditional optimality conditions (marginal cost pricing) do not necessarily apply in all other sectors. In essence, the results obtained in a "second best" analysis may contradict the intuition based on a first best analysis. In this case, the optimality of transit subsidies derive from the underpricing of roads. Since automobile travel creates negative effects in terms of congestion and pollution, reducing auto travel demand will have economic benefits as long as the marginal cost of inducing a driver to take transit is less than the marginal social cost imposed by driving.

A Framework for Optimal Transit Subsidies In the 1970s, Glaister and Lewis developed a method for calculating the optimal (i.e., welfare-maximizing) fare structure for public transit when road pricing is not a viable option. The author's method has withstood scrutiny since then and remains a standard reference in the literature on economic welfare and public transit.

The argument is that since private vehicle users are charged less than their marginal social cost[181] of driving, particularly during congested conditions, there is an economic rationale for pricing public transport below it's marginal cost to induce drivers to switch to public transit. This conclusion rests on the actual marginal social costs of driving and on the ability of reductions in transit fares to attract travelers away from road use.

This paper adopts the Glaister and Lewis method and presents a new application of the model using risk analysis techniques. These techniques account for the inherent uncertainty in many of the model's inputs. The model also incorporates recent advances in the ability to determine the marginal social costs of automobile travel.

The Framework A detailed derivation of the model is presented in technical form in Annex 3.2. The central element of the "second best" theory is the congestion to all traffic caused by an additional private vehicle during peak-period automobile travel. Under congested conditions, each additional automobile in the transportation network imposes high costs in terms of congestion (lost time and inconvenience) upon all vehicles in the affected transportation network. Because each additional vehicle does not pay for the costs imposed on other transportation network users, it is referred to as an external cost.

The theory incorporates this external congestion effect to show that, when the various transportation modes are substitutes for each other, transit subsidies can be an efficient (i.e., welfare-maximizing) policy response to congestion in the absence of correct road pricing.

Implications of the Second Best Result The degree to which transit subsidization creates net benefits depends on the ability of fare reductions for transit to attract travelers away from road use and, in addition, on volumes and shares on the various modes, and the actual marginal social costs of automobile travel.

The "second-best" transit subsidy ensures that the price of transit, relative to automobile travel is optimal, or welfare maximizing. This results in the most efficient distribution of traffic across modes and ensures that transit and road users benefit mutually.

The theory suggests that the optimal fares on public transit modes are below their marginal costs in both peak and off-peak periods. Subsidies are justified in the peak period because lower fares induce mode switching from auto to public transit which reduces traffic congestion. Subsidies during off-peak periods are justified because they induce people to travel in the off-peak period, reducing peak-period congestion.

Model Structure and Data The second-best solution to the transit subsidy problem uses an analytical method well grounded in economic theory to combine data on (i) the marginal social cost of automobile use; (ii) the marginal social cost of transit (bus and rail separately); (iii) the responsiveness of the demand for bus and rail service to fare changes; and (iv) the responsiveness of the demand for auto use with respect to bus and rail fares.

Table 3.13 presents the elasticities that were used in the analysis. These elasticities were estimated by Glaister and Lewis (1978) and represent conservative estimates compared to many studies in the literature.[182] They measure the responsiveness of travelers using a current mode of transportation (say, a car) to changes in the prices of other modes (say, a bus). For example, an elasticity of demand of 0.025 for auto travel with respect to bus fares would show that an increase of 100 percent in bus fares would lead to a 2.5 percent increase in the number of auto travelers.

Table 3.13 Elasticities Adopted for Analysis

Mode		Bus		Rail		Auto
	Period	Peak	Off Peak	Peak	Off Peak	Peak
Bus	Peak	-0.35	0.04	0.14	0.01	0.025
	Off Peak	0.029	-0.87	0.009	0.28	0.0016
Rail	Peak	0.143	0.013	-0.30	0.05	0.056
	Off Peak	.008	0.28	0.018	-0.75	0.0034

Source: Glaister and Lewis (1978).

For higher elasticities between transit and auto use, drivers are more responsive to fare reductions, which would relieve congestion more easily. The elasticities used in this model would seem low, and the resulting benefits from congestion relief would be similarly conservative. The responsiveness of automobile drivers to transit fares is estimated to be extremely small. But even small degrees of automobile driver responsiveness to transit fares can translate into significant levels of efficient subsidies.

The model is calculated using operating costs for seven metropolitan areas. It includes capital costs in the amount used only for current system renovation, maintenance and improvement, omitting all capital spending on new systems and segments.

Capital Subsidies Governments typically provide nearly all of the capital improvement budgets for transit agencies. This issue is problematic for the calculation of the optimal subsidies because the portion of the capital budget that is directed toward expanding capacity or extending a transit line will have demand impacts in the future that will not show up in current year data.

This analysis takes the current transportation infrastructure as given. A major portion of capital expenditures represent infrastructure expansions which become usable in the future. This analysis is a tool for determining the optimal transit subsidies, given the current infrastructure, and does not give any guidance regarding the wisdom or impact of new capital expenditures. New investments should be subjected to rigorous cost-benefit analysis to determine whether they yield adequate net benefits.

Some portion of capital expenditures are used for general maintenance of the current infrastructure. This amount should be considered part of current operating subsidies since this spending is required to maintain the transportation system in its current state (see Table 3.14). Unfortunately, current data on the precise distribution of capital costs does not exist for most transit agencies.

Table 3.14 Transit Capital Funds Applied

	Percent of Capital Funds Applied (1991)			Capital Spending (Million 1993)	Annual Pass Miles (Million 1993)		Cents Per Pass. Mile	
	Bus	Existing Rail	New Rail		Bus	Rail	Bus	Rail
Los Angeles	20.2	7.6	72.2	$350.7	1916.1	145	4	18
Wash. DC	7.2	2.9	89.9	$261.5	603.4	968	3	1
Chicago	50.2	47.0	2.8	$440.5	1031.3	2,248	21	9
Boston	NA	92.3	7.7	$235.4	254.5	1,018	0	21
New York	15.9	83.6	0.5	$1,405.2	2152.1	10,23	10	11
Philadelphia	22.8	77.2	NA	$253.6	471.3	775	12	25
San Francisco	34.5	65.5	NA	$218.6	595.2	1,048	13	14

Source: Hickling Lewis Brod Economics Estimates and U.S. Department of Transportation, *National Transit Database.*

Capital funds applied to the current infrastructure are estimated, based on data from Federal Transit Administration National Transit Database (NTD). Data regarding the distribution of capital costs were, in fact, collected by FTA up to 1991. By calculating the percentage of capital funds applied to existing infrastructure in 1991, and applying this percentage to capital funding levels in 1993, an estimate of the capital funds applied to existing infrastructure is generated.

The solution to estimating capital funds applied to current infrastructure, while not perfect, should produce acceptable estimates of the distribution of capital funds by transit agency and by purpose. The implicit assumption in the calculation is that the composition of capital spending in 1993 is approximately the same as 1991. Including capital costs is important to derive the correct value for the marginal cost of transit service. Especially for older transit systems, maintenance and rehabilitation of the existing infrastructure can demand large levels of capital spending (see Table 3.14). The "second best" subsidy calculation should and does account for the true costs of providing transit service.

Results In order to account for uncertainty in model inputs, the second best model was applied using risk analysis techniques. Rather than rely on point estimates for model inputs, the risk analysis approach uses ranges for all model inputs to account for uncertainty. The risk analysis results provide policy makers with a quantitative basis to make decisions which fully accounts for uncertainty.

Model Results The optimal fares were calculated for seven large United States metropolitan areas and for the nation as a whole. The results appear in Table 3.15. Among other things, they present the subsidies that should be given to transit authorities in order to enable them to charge a welfare-maximizing or "second-best" fare given their current operating costs and capital expenditures on current infrastructure. Since the second-best model developed in this paper assumes that the infrastructure is given, the correct subsidy predicted by this model should be interpreted as an operating and maintenance subsidy.

Conclusion The results of our analysis confirm that subsidizing public transportation can be the best public policy approach to utilizing the transportation infrastructure in the absence of road pricing. This analysis suggests the optimal subsidy levels in major metropolitan areas is a significant portion of total transit costs (see Table 3.15) based solely on transit's congestion management benefits.[183] The optimal subsidy exceeds the current subsidy in these systems and the nation as a whole.

The central result of this analysis is that all travelers on the highways and transit systems could be made better off by increasing transit subsidy levels in several major metropolitan areas.This conclusion is, of course, a "second best" argument where alternative policies toward congestion management, such as road pricing, are deemed not feasible for political or practical reasons. The model presented here would suggest that the optimal subsidy would decline as the fees that drivers face approach the marginal social cost of automobile travel. The "first best" approach would be to remove all price distortions in the transportation market. The "second best" approach provides a means of improving the allocation of resources among the available transportation modes when the first best approach is unavailable. The results of our analysis confirm that subsidizing existing public transportation can cause congestion relief benefits. This conclusion is, of course, in a "second-best" world where road pricing is neither politically nor institutionally feasible.

Table 3.15 Second-Best Subsidy Results

Urbanized Area	Optimal Operating[a] Subsidy ($Millions)	Optimal Subsidy as Percent of 1993 Operating[a] Costs	Optimal Subsidy as Percent of 1993 Total Costs[b]
Boston	$812.10	101.2	99.4
Chicago	$1,421.40	91.2	89.1
Los Angeles	$1,123.10	116.3	76.0
Philadelphia	$711.65	85.6	85.6
New York.	$6,290.60	95.9	95.8
San Fran.	$788.60	73.1	56.0
Wash. DC	$665.26	98.3	71.3
National Est.	$15,995.9	100.1	78.9

[a]Operating subsidies and costs include capital expenditures applied to the current transit system.
[b]Total costs include operating costs and all capital expenditures on current infrastructure and new starts.

Source: Hickling Lewis Brod Economics Estimates and U.S. Department of Transportation, *National Transit Database*.

Annex 3.1 Corridor Study Technical Annex

The Economic Theory of Convergence

The theory presented here follows the standard model from public economics of utility maximization under a budget constraint with an external effect. Consider an individual who derives utility from consuming z units per week of a basket of commodities. In order to generate the income required to purchase the consumption good, he (or she) must take x trips per week (say, five inbound and five outbound) from a residential area to a central business district. The individual derives disutility, however, from the amount of time spent traveling. While disutility may be derived differently from different types of travel time (i.e., driving, riding, walking, waiting in congestion, etc.) for simplicity, the individual is assumed to be indifferent between travel times of different types. The individual can choose to travel by one of two modes, highway or high-capacity transit, each of which has a money price associated with the trip.

If there are I individuals, the utility maximization problem of the ith individual is expressed as:

$$\max \; u^i(z\,,\,t)$$

$$s.t. \;\; x^i_1 P_1 \;+\; x^i_2 P_2 \;+\; z \;\leq\; y^i \qquad \text{eq. 1}$$

where t represents time spent commuting, $x_1{}^i$ and $x_2{}^i$ are the number of trips taken by the highway and the transit modes, respectively. The prices P_1 and P_2 are the money cost of a trip by each mode, y^i is the individual's income and z is a numeraire representing all other goods.

The utility function is assumed to be continuous and twice differentiable, having the following properties:

$$u^i_z > 0\,,\; u^i_{zz} < 0\,,\; u^i_t < 0 \text{ and } u^i_{tt} < 0 \qquad \text{eq. 2}$$

The conditions on z are the regular strong concavity conditions for consumption goods. Time spent traveling is a "bad" which the individuals would be willing to pay to avoid. Concavity with respect to t implies an increasing marginal disutility—the more time spent traveling, the greater the disutility from additional travel time.

The individual must allocate his total number of trips among the two modes:

$$x^i \;=\; x^i_1 \;+\; x^i_2 \qquad \text{eq. 3}$$

The trip time by the highway mode is an increasing function of the number of trips taken by all travelers:

$$t_1 = d + a\left(\frac{X_1}{v - X_1}\right)^b \quad where\ X_1 = \sum_{i=1}^{I} x_1^i \qquad \text{eq. 4}$$

d represents an uncongested, "freeflow" travel time and v represents the capacity constraint of the highways, i.e., the upper bound on the number of trips which could be taken by highway which would result in gridlock and an infinite trip time. a and b are structural parameters reflecting the speed-volume relationship of the highway network. X_1 represents the total number of trips by all travelers via the highway mode.

The high-capacity transit mode is assumed to be completely unaffected by additional trips and the trip time is a fixed value:

$$t_2 = c \qquad \text{eq. 5}$$

The transit mode is assumed to be a "high-speed" mode where the linehaul segment of a journey is rapid relative to, say, the expressway segment of a highway journey thus compensating for slower speeds accessing the high-capacity mode including walk and wait times.

The absence of an external effect from additional riders on the high-capacity mode is expressed by eq. 5. Of course, crowding on transit results in some riders standing and other inconvenience. However, the key operational assumption is that travel times on the high speed mode are unaffected by changing volumes of passengers which corresponds to the actual scheduling practice in rail transit systems.

Time spent commuting is given by the sum of trips weighted by the average time per trip. The ith commuter's total travel time is given by:

$$t^i = x_1^i t_1 + x_2^i t_2 \qquad \text{eq. 6}$$

The total trip time by the individual can be expressed as a function of the number of highway trips by substituting eq. 4 and eq. 5 into eq. 6:

$$t^i(x_1^i) = x^i c + (d - c) + a\left(\frac{X_1}{v - X_1}\right)^b x_1^i \qquad \text{eq. 7}$$

The first order conditions of utility maximization are given by:

$$P_1 - P_2 = \frac{u_{x_1}^i}{u_z^i} = \frac{u_t^i}{u_z^i}\frac{\partial t^i}{\partial x_1^i} \qquad \text{eq. 8}$$

Where:

$$\frac{\partial t^i}{\partial x_1^i} = (d - c) + a\left[\frac{X_1}{v - X_1}\right]^b \left[1 + \frac{x_1^i \; b \; v}{(v - X_1) \; X_1}\right]$$

$$= t_1 - t_2 + \left[\frac{abv}{v - X_1}\right]\left[\frac{x_1^i}{X_1}\right]\left[\frac{X_1}{v - X_1}\right]^b$$

eq. 9

Some individuals will maximize utility by choosing all trips by one mode or another. However, some individuals will find their optimum allocation of trips by a mix of trips on both modes. These are "casual" switchers—that is, their circumstances or preferences do not lock them into a particular mode—and they correspond to the modal explorers discussed in the introduction. Note that eq.9 can be re-arranged to give:

$$\left[(P_1 - P_2)\frac{u_z^i}{u_t^i}\right] - \left[\left(\frac{abv}{v - X_1}\right)\left(\frac{x_1^i}{X_1}\right)\left(\frac{X_1}{v - X_1}\right)^b\right] = t_1 - t_2$$

eq. 10

or, the condition under which door-to-door journey times across modes will be equal is given by:

$$\left[(P_1 - P_2)\frac{u_z^i}{u_t^i}\right] = \left[\left(\frac{abv}{v - X_1}\right)\left(\frac{x_1^i}{X_1}\right)\left(\frac{X_1}{v - X_1}\right)^b\right]$$

eq. 11

Condition 11 tells us what combinations of prices, congestion, personal preferences and highway speed-flow relationship will result in equal travel times. However, it can be readily shown that under the assumptions described above—especially the assumption of an growing marginal disutility with respect to travel time—that with sufficient levels of congestion both the left and right hand sides of eq. 11 approach zero.

What happens under congested conditions? The left hand side tends to zero due to the growing marginal disutility from increased travel time (also, the left hand side approaches zero with increasing income—the individual becomes indifferent to the price differential as trip cost consumes a smaller portion of his income). The theory also implies that congestion pricing will be less effective as congestion becomes more severe. It can be readily shown that if u_t^i is not bounded then for any combination of prices and capacity equation parameters, and for any small value □>0 , there is a level of congestion (number of total trips) sufficiently large such that:

$$|t_1 - t_2| < \varepsilon$$

eq. 12

Annex 3.2 "Second Best" Technical Annex

Throughout this paper and in keeping with the original Glaister–Lewis framework, we use the following indices:

1 : peak hour private vehicle	4 : no bus
2 : no private vehicle	5 : peak hour rail
3 : peak hour bus	6 : no rail

Elasticities The cross-price elasticity of demand for transportation services on mode i with respect to prices on mode j will be given by the standard equation for price elasticity as follows:

$$\eta^i_j = \frac{p_j}{X^i}\frac{\partial X^i}{\partial p_j} = \frac{p_j}{X^i}X^i_j, \quad i,j = 1,\ldots,6 \qquad \text{eq. 13}$$

Where

p_i are the prices on mode i in \$ per passenger mile, and
X^i are the demands on mode i in passenger miles.

If cross-price elasticities of auto travel with respect to public transit fares are estimated to be zero, implying that there is no way of persuading automobile users to switch to buses or rail transit regardless of price, then the Glaister-Lewis model would predict that transit fares should be set at the marginal cost of delivering service. If these elasticities do not equal zero, some level of transit subsidy will be efficient in the absence of road pricing.

Deriving the Equation System Glaister-Lewis conceived of the consumer's problem as a maximization of the consumer's expenditure function less the operating costs of the various public transit modes. The maximization, following the Glaister-Lewis paper, can be expressed as follows:

$$\begin{aligned}
&\max_{p_3,p_4,{}_5,p_6} \{G(\alpha_3,\alpha_4,\alpha_5,\alpha_6,X^1(\alpha_3,\ldots,\alpha_6),X^3(\alpha_3,\ldots,\alpha_6),\hat{p},u) \\
&-G(p_3,p_4,p_5,p_6,X^1(p_3,\ldots,p_6),X^3(p_3,\ldots,p_6),\hat{p},u) \\
&-[C^3(X^1,X^3)-p_3X^3]-[C^4(X^4)-p_4X^4]-[C^5(X^5)-p_5X^5] \\
&-[C^6(X^6)-p_6X^6]\}
\end{aligned} \qquad \text{eq. 14}$$

Where

$G(p, X^1, X^3, \hat{p}, u)$ is the expenditure function aggregated across individuals,

X^i is the traffic level for mode i

$(p_3,...,p_6)$ is a vector of transit fares,

$\hat{p}$ is a vector of all other prices including p_1 and p_2.,

u is a vector of constant utility levels, and

C^i are the operating costs of the public transit modes.

The expenditure function, representing the long run demand responses to prices, depends on peak car and bus traffic levels because of the negative effects of congestion on consumer utility. This relationship implies that for a given vector of prices, an increase in peak traffic requires a compensating increase in income to maintain the previous level of consumer utility. This relationship is known as the compensating variation and is given by the difference between expenditure function evaluated at the "reference" prices α_i and a lower set of prices p_i. The compensating variation is the amount of money that would be required to compensate for an increase from $p_3,...,p_6$ to $\alpha_1,..., \alpha_6$, where the α_i's represent higher peak-hour congestion levels than the p's.

The other terms within the [ ] are the operating subsidies required for the peak and off-peak bus and rail transit services. The compensating variation and the public transit fare revenues (p_iX^i) represent consumers' total willingness-to-pay from which the transit systems' operating expenses ($C^i(X^i)$) must be subtracted.

eq. 14 is differentiated with respect to p_3, p_4, p_5, and p_6. Differentiating eq. 14 with respect to p_3 yields one of four first-order conditions for a maximum as follows:

$$-\frac{\partial G}{\partial p_3}-\frac{\partial G}{\partial X^1}X_3^1-\frac{\partial G}{\partial X^3}X_3^3-C_1^3X_3^1-C_3^3X_3^3+X^3$$
$$+p_3X_3^3-C_4^4X_3^4+p_4X_3^4-C_5^5X_3^5+p_5X_3^5-C_6^6X_3^6+p_6X_3^6=0$$

eq. 15

Similar expressions are obtained from differentiating with respect to p_4, p_5 and p_6. Using the following properties and definitions:

$$\frac{\partial G}{\partial p_i} = \sum_h \frac{\partial g_h}{\partial p_i} = \sum_h x_h^i = X^i, \quad S_1 = \frac{\partial G}{\partial X^1} + \frac{\partial C^3}{\partial X^1} \text{ and } S_3 = \frac{\partial G}{\partial X^3} + \frac{\partial C^3}{\partial X^3}.$$

eq. 16

where S_1 is the marginal social cost of peak automobile travel per passenger mile and S_3 represents the marginal social cost of peak bus travel per passenger mile. Substituting these expressions into the first order condition expressed in eq. 15 and collecting terms results in the following expression:

$$(p_3 - S_3)X_3^3 + (p_4 - C_4^4)X_3^4 + (p_5 - C_5^5)X_3^5 + (p_6 - C_6^6)X_3^6 = S_1 X_3^1$$

eq. 17

Similar expressions are obtained from the other three first-order conditions after substituting and rearranging terms.

Optimal Fares and Subsidies System of Equations The equation system that allows the calculation of the "second-best" optimal fare derives from the four first-order conditions for the maximization problem in eq. 14. The first order conditions, after converting to elasticity form, reduce to:

$$\left[\eta_3^3(p_3 - S_3)X^3 + \eta_3^4(p_4 - C_4^4)X^4 + \eta_3^5(p_5 - C_5^5)X^5 + \eta_3^6(p_6 - C_6^6)X^6\right]\frac{1}{S_1 X^1} = \eta_3^1$$

$$\left[\eta_4^3(p_3 - S_3)X^3 + \eta_4^4(p_4 - C_4^4)X^4 + \eta_4^5(p_5 - C_5^5)X^5 + \eta_4^6(p_6 - C_6^6)X^6\right]\frac{1}{S_1 X^1} = \eta_4^1$$

$$\left[\eta_5^3(p_3 - S_3)X^3 + \eta_5^4(p_4 - C_4^4)X^4 + \eta_5^5(p_5 - C_5^5)X^5 + \eta_5^6(p_6 - C_6^6)X^6\right]\frac{1}{S_1 X^1} = \eta_5^1$$

$$\left[\eta_6^3(p_3 - S_3)X^3 + \eta_{63}^4(p_4 - C_4^4)X^4 + \eta_6^5(p_5 - C_5^5)X^5 + \eta_6^6(p_6 - C_6^6)X^6\right]\frac{1}{S_1 X^1} = \eta_6^1$$

eq. 18

This system of equations fully identifies the optimal transit pricing structure in the absence of road pricing. This system can be written in matrix notation as follows:

$$\begin{bmatrix} \eta_3^3 & \eta_3^4 & \eta_3^5 & \eta_3^6 \\ \eta_4^3 & \eta_4^4 & \eta_4^5 & \eta_4^6 \\ \eta_5^3 & \eta_5^4 & \eta_5^5 & \eta_5^6 \\ \eta_6^3 & \eta_6^4 & \eta_6^5 & \eta_6^6 \end{bmatrix} \begin{bmatrix} (p_3 - S_3)X^3 \\ (p_4 - C^4)X^4 \\ (p_5 - C^5)X^5 \\ (p_6 - C^6)X^6 \end{bmatrix} \frac{1}{S_1 X^1} = \begin{bmatrix} \eta_3^1 \\ \eta_4^1 \\ \eta_5^1 \\ \eta_6^1 \end{bmatrix}$$

eq. 19

Solving the System The preceding system of equations is a set of four equations with four unknowns which is solvable using linear algebra techniques. The object of this project is to determine the values of the p's in the equations from which the optimal subsidy levels can be calculated.

This model can be applied to transportation systems with automobile, bus and rail modes. When rail is not available, the system of equations reduces to two equations with two unknowns as follows:

$$\left[\eta_3^3(p_3 - S_3)X^3 + \eta_3^4(p_4 - C_4^4)X^4\right]\frac{1}{S_1 X^1} = \eta_3^1$$

$$\left[\eta_4^3(p_3 - S_3)X^3 + \eta_4^4(p_4 - C_4^4)X^4\right]\frac{1}{S_1 X^1} = \eta_4^1$$

eq. 20

The difficulty in solving the system increases rapidly with the number of modes and periods under consideration. This system can be expressed as a linear system and solved using matrix inversion. This system does not provide explicit solutions for the optimal fares, but these can be calculated using some assumed functional forms for the demand and cost functions.

Applying matrix inversion and solving for the auto-bus-rail system will result in a numerical solution for the column vector (the auto-bus case results in numerical solutions for the first two elements of the following vector):

$$\begin{bmatrix} (p_3 - S_3)X^3 \\ (p_4 - C^4)X^4 \\ (p_5 - C^5)X^5 \\ (p_6 - C^6)X^6 \end{bmatrix} \frac{1}{S_1 X^1}$$

eq. 21

Estimates for S_i, C^i, and X^i can be obtained or estimated from secondary sources and using standard functional forms for the cost and demand functions. The p_i's can then be determined by simple algebra.

Data for the "Second Best" Model Secondary data sources provide a set of parameters with which to calculate the optimal subsidies for a set of transit systems. The original Glaister-Lewis paper relied on a set of secondary sources augmented by sensitivity analysis to account for uncertainty in some of their variables. This application of the methodology is augmented by risk analysis to account for uncertainty surrounding the values chosen to estimate the equation system.

Input Requirements In order to estimate p_i, the second-best price for each mode on-peak and off, all other variables in the system, presented in Annex 2, must be estimated or identified. The inputs needed to solve this system are presented in Table 3.16 below.

Table 3.16 Input Requirements for Second Best Model

Variable	Description
η^i_j	Cross-price elasticity of demand[184] for mode i with respect to prices on mode j for $i,j \in \{1,\ldots 6\}$.
S_1	Marginal social cost of private vehicle travel per passenger mile during peak hour.
S_3	Marginal social cost of bus travel per passenger mile during peak hour.
C^4	Operating costs of the off-peak bus transit per passenger mile.
C^5	Operating costs of the peak rail transit per passenger mile.
C^6	Operating costs of the off-peak rail transit per passenger mile.
X^1	Demand for peak auto travel in passenger miles.
X^3	Demand for peak bus travel in passenger miles.
X^4	Demand for off-peak bus travel in passenger miles.
X^5	Demand for peak rail travel in passenger miles.
X^6	Demand for off-peak rail travel in passenger miles.

Data Sources and Tables Elasticity figures are not readily available, but must typically be estimated econometrically. In this case, we have chosen to adopt the Glaister-Lewis estimates for the elasticities. These elasticities were derived from the economics literature and from previous work by David Lewis.[185]

Table 3.17 shows the adopted elasticity estimates for the optimal subsidy model. As can be seen in the table, all entries are non-zero, but many are very small. The marginal social cost variables need to be determined. In each case, the marginal costs of travel need to be increasing

Table 3.17 Adopted Elasticities for Optimal Subsidy

		Bus		Rail		Auto
		Peak	Off Peak	Peak	Off Peak	Peak
Bus	Peak	-0.35	0.04	0.14	0.01	0.025
	Off Peak	0.029	-0.87	0.009	0.28	0.0016
Rail	Peak	0.143	0.013	-0.30	0.05	0.056
	Off Peak	.008	0.28	0.018	-0.75	0.0034

Source: Glaister and Lewis (1978)

with traffic to account for the impact of congestion on the cost of travel. The marginal social cost of bus service is a combination of congestion costs and system operating costs. The estimates used in this analysis were obtained from work by Dr. Herbert Mohring and David Anderson (work in progress) combined with operating costs for the bus system. The adopted marginal social cost estimates derived from Dr. Mohring are in Table 3.18.

Table 3.18 Marginal Social Cost Estimates

Variable	Adopted Value
Marginal Social Cost of Car Travel	\$0.49/passenger mile
Marginal Social Cost of Bus Travel	\$0.40/passenger mile

Source: Mohring, H. and D. Anderson, in progress and Hickling Lewis Brod Economics estimates.

Operating costs for the public transit system are based on available system operating cost data found in the National Transit Database Transit Profiles.[186] There are assumed to be no congestion costs off peak. The only justification, according to this modeling structure, for subsidizing off-peak transit service is to shift travelers from peak to off-peak period travel, reducing peak congestion.

For these model runs, demand estimates for 1993 were derived from Section 15 data and from the Texas Transportation Institute.[187] A log-linear functional form was assumed for the demand function based on price of service and the elasticity estimates in Table 3.17. Other functional forms can be used in this model depending on the preference of the analyst. The only limitation on the adoption of functional demand equations, are that functional forms must contain only variables determined or estimated in this model or known values to maintain the solvable "four equations/four unknowns" equation system structure.

Notes

117 Lawrence Sanders, *The Second Deadly Sin,* (New York: Berkley Books, 1977), p. 126.

118 Joseph E. Stiglitz, *Economies of the Public Sector,* (New York: W.W. Norton & Company, 1986), p.107.

119 Kenneth J. Button, *Transport Economics–2nd Edition*, (Vermont: Edward Elgar Publ Ltd., 1993), p. 110.

120 Richard Arnott and Kenneth Small (1994) "The Economics of Traffic Congestion–Rush-hour driving strategies that maximize an individuals driver's convenience may contribute to overall congestion", *American Scientist*, Vol. 82, p. 451.

121 Terry Moore & Paul Thorsnes, "The Transportation/Land Use Connection: A Framework for Practical Policy", *Report Number 448/449*, (Washington D.C.: American Planning Association, 1994), p. 31.

122 Ibid.

123 Anthony Downs, *Stuck in Traffic–Coping with Peak-Hour Traffic Congestion*, (Washington D.C.: The Brookings Institution, 1992), p. 7.

124 D.L. Schrank, S.M. Turner and T.J. Lomax, *Estimates of Urban Roadway Congestion–1990*, (Washington, D.C.: United States DOT, 1993), p. 51.

125 Bureau of the Census, *Statistical Abstract of the United States 1993. 113th Edition.* Washington, D.C.: U.S. Department of Commerce, p. 612, Table No.1009.

126 Holtzclaw (1994), Button (1993), Downs (1992), Mogridge (1990).

127 Shrank et al (1993) p. 25.

128 Texas Transportation Institute.

129 Martin J.H. Mogridge, loc. cit.

130 Shrank et al, op. cit., p. 24.

131 Downs, op. cit., p. 20.

132 Fairlie, op. cit., p. 214.

133 Martin J.H. Mogridge, op. cit., p. 281.

134 Ibid. p. 212

135 Elmer Johnson, "Collision Course: Can Cities Avoid a Transportation Pileup?" *American City and Country*. Vol.109, Issue 3, (March 1994). pp. 43-45.

136 John Holtzclaw, *Using Residential Patterns and Transit to Decrease Auto Dependence and Costs*, (San Francisco: Natural Resources Defense Council, 1994), p. 4.

137 Johnson, op. cit., p. 54

138 John R. Meyer and Jose A. Gomez-Ibanez, *Autos Transit and Cities*, (London: Harvard University Press, 1981), p. 211.

139 Ibid. p. 211

140 Downs, op. cit., p. 34.

141 Mogridge, op. cit., p. 282.

142 R.J. Smeed and J.G. Wardrop, "An Exploratory Comparison of the Advantages of Cars and Buses for Travel in Urban Areas", *Journal of the Institute of Transport*, Vol.30, No.9, 1964), pp. 331-315.

143 C.H. Sharp, "The Choice Between Cars and Buses on Urban Roads", *Journal of Transport Economics and Policy*, Vol.1, (1967), pp. 104-11.

144 P.B. Goodwin, "Car and Bus Journeys to and from Central London in Peak Hours", *Traffic Engineering and Control*, Vol.11, (1969). pp.376-78.

145 F.V. Webster and R.H. Oldfield, *A Theoretical Study of Bus and Car Travel in Central London*, (Crowthorne, Berks, U.K.: Transport and Road Research Laboratory, 1972).

146 John Kain, "Impacts of Congestion Pricing on Transit and Carpool Demand and Supply", *Curbing Gridlock: Peak-Period Fees to Relieve Traffic Congestion*, Vol.2, (Washington, D.C.: National Research Council, 1994), p. 539.

147 A. Downs, "The Law of Peak-hour Expressway Congestion", *Traffic Quarterly*, Vol.16, 1962), pp.393-409.

148 Mogridge, op. cit., p. 212.

149 W. Suchorzewski, "Principles and Applicability of the Integrated Transportation System", *Proceedings of the UN-ECE Seminar on the Role of Transportation in Urban Planning, Development and Environment*, (Muenchen, Germany: 1973).

150 Downs, op. cit., p. 27.

151 Ibid.

152 Mogridge, op. cit., p. 193.

153 Ibid., p. 164.

154 A.C. Pigou, *The Economics of Welfare*, (London: Macmillan, 1920).

155 Button, op. cit., p. 165.

156 F.Y. Edgeworth, *Papers Relating to Political Economy*, (London., 1925).

157 H. Hotelling, Edgeworth's Taxation Paradox and the Nature of Supply and Demand Functions, *Journal of Political Economy*, Vol. 40, No. 5, 1932), pp. 577-616.

158 Mogridge, op. cit., p. 10.

159 Moore and Thorsnes (1994) p. 106.

160 Mogridge, (1990) p. 283.

161 C. Buchanan, Traffic in Towns: A Study of the Long-Term Problems of Traffic in Urban Areas, (London, 1963).

162 Mogridge, (1990) p. 151.

163 Automobile Association of America, *Your Driving Costs*, 1995 Edition, (Washington, D.C.).

164 APTA, *Transit Fact Book*, 1994-1995 Edition, (Washington, D.C.: American Public Transit Association) (revenue figures for the year 1993).

165 James J. MacKenzie, Roger C. Dower and Donald D.T. Chen, *The Going Rate: What It Really Costs to Drive* (Washington, D.C.: World Resources Institute, 1992) (Highway cost statistics are for the year 1989).

166 Peter Miller and John Moffet, *The Price of Mobility*, (San Francisco: Natural Resources Defense Council, 1993) (Subsidy estimates are for the year 1990).

167 Willingness to pay is the amount travelers would be willing to pay for their transportation, including out-of-pocket and time costs. This amount usually exceeds the amount they actually do pay and the difference between the two is a measure of benefits to the consumer.

168 Transportation Research Board, *Record,* No.: 332 (Washington, D.C.: National Research Council, 1996).

169 Brod, D. "Accounting for Multi-Modal System Performance in Benefit--Cost of Transit Investment", Forthcoming, Transportation Research Record, (Washington, D.C.: National Research Council, 1996).

170 Mogridge, M.J.H. (1990).

171 Downs, (1992).

172 National Academy of Sciences, *Curbing Gridlock: Peak Period Fees to Relieve Traffic Congestion*, Special Report 242, (Washington, D.C.: National Research Council, 1994).

173 Hickling Lewis Brod Inc., "Unsticking Traffic: When Transit Works and Why", An FTA Policy Paper, October, 1994.

174 Also in economic terms, making consumers better off is also expressed as increasing their welfare. When consumers can't be made any better off, then their welfare is said to be maximized.

175 Laffont, Jean –Jacques, *Fundamentals of Public Economics*. (Cambridge: The MIT Press, 1989), p.167.

176 This result depends on the fact that mass transit causes less congestion than does road travel.

177 Glaister, S. and D. Lewis. "An Integrated Fares Policy for Transport in London", *Journal of Public Economics*, Vol. 9, 1978, pp. 341-55.

178 Ibid.

179 Cross-price elasticities refer to the responsiveness of consumers to price changes on alternative travel modes.

180 Mohring, H. and D. Anderson, "Congestion Costs and Congestion Pricing", in progress.

181 The marginal social cost is the social cost associated with newest road user.

182 Button, Kenneth, *Transportation Economics*, (Cambridge: University Press, 1993), p. 47.

183 Transit costs include many costs not related to congestion management. Examples of these are paratransit services and low-cost mobility programs, among others.

184 The cross-price elasticity of demand for commodity I with respect to the price of commodity j is the responsiveness of the consumers' demand for commodity I (in percentage terms) to a change in the price of commodity j (also in percentage terms).

185 Lewis, David, "Estimating the Influence of Public Policy on Road Traffic Levels in Greater London", *Journal of Transport Economics and Policy*, (May, 1977).

186 *Transit Profiles: Agencies in Urbanized Areas Exceeding 200,000 Pop*ulation, (Washington, D.C.: United States Department of Transportation, 1995).

187 Texas Transportation Institute, *Estimates of Urban Roadway Congestion* - 1990, Research Report 1131-5, (Washington, D.C.: Transportation Research Board, 1993).

4 The Low Cost Mobility Benefits of Transit

Merit Goods and Low Cost Mobility

The concept of merit goods was developed by Musgrave (1959), as a category of goods for which the public makes a judgment that consumption is either "desirable" or "undesirable".[188] Merit goods are judged differently by public (societal) and private interests creating a "valuation gap" and a tendency to under-consume "desirable" goods. This valuation gap has been used as the justification for public sector involvement, either through direct provision or through subsidies.

The nature of merit goods suggests that inducements for the consumption of "desirable" goods and services are based on a policy judgment which in turn is based on collective preferences. As an example, low-cost housing and mobility are in some cases considered merit goods because they meet the basic needs of society.[189] Vaccinations and education are also viewed as merit goods because they promote healthy and active participation in society.

The social nature of the merit good framework does not alter the real economic consequences of encouraging the consumption of a particular type of good. For instance, the value of education is undisputed as the fundamental input for the development of human capital. And, promoting the long-term health of society through vaccination programs is an economically desirable outcome from a macro-economic point of view. However, if left to individual behavior, there is little doubt that such goods would be under-consumed and GDP would be diluted accordingly.

Although the notion of "merit goods" has a "elitist" ring to it, the vast majority of merit goods are deeply imbedded in modern society and attract little controversy. Most public sector activities have become part of the fabric of everyday life so that the public expenditure for them is taken for granted. Obvious cases are public education, fire protection, road building, sidewalk maintenance, law enforcement, and environmental protection.

Each one of these examples, over time, has evolved to include collateral merit goods. We expect our schools, for example, to teach much more than reading, writing and arithmetic. We expect our firefighters to accept serious personal risks and we compensate them accordingly.

One might say that the particular mix of "merit goods" that attracts enduring public support in a given community is a fair reflection of its political culture, even of the civilization itself. The legacy of a society or of a political order is a merit good.[190]

The definition of low cost mobility as a merit good is conceived in both a social and an economic framework. Individuals with access to basic mobility enjoy the benefits of social interaction, entertainment and education which in turn influences their contribution to society and the economy. And, basic mobility extends the opportunities for employment to individuals who, without this access may otherwise be unemployable. These characteristics of basic mobility, taken together, make a positive contribution to the economy.

The concept of merit goods as it applies to low-cost mobility has a long tradition in the transportation economics literature. Many studies suggest that mobility is best characterized as an entitlement to a minimum standard of transport provision. As Mayer *et al* (1973) state "[m]obility deprivation should be considered in the same way as other forms of deprivation, such as housing, education and employment....policies [should] cater for the personal mobility of everyone".[191] Mayer *et al* (1973) also assert;

> "The old, infirm and children are obvious examples where irrespective of income, effective demand may be felt an inadequate basis upon which to allocate transport resources....these groups are in need of *adequate and inexpensive public transport services and the normal market mechanism is inadequate in this respect*".[192]

Button (1993) maintains that mobility should not be allocated on the basis of effective demand but rather on a concept of need; "closely concerned with the notion of merit goods–that is, needs 'considered so meritorious that their satisfaction is provided for through the public budget over and above what is provided for through the market and paid for by private buyers".[193]

The concept of merit goods suggests an argument for the subsidization or direct provision of these goods by government agencies. Button (1993) asserts that there are justifications for providing services that experience financial losses if external benefits, in addition to those directly attributable to transportation, exist: "[T]he service may, for instance, have a strategic value or it may be deemed a 'merit good' which society ought to provide to enable isolated communities to continue in existence".[194] In most cases, a combination of direct provision and direct subsidies are used to support the provision of basic mobility. The argument that such provisions should be present is further sustained by the fact that "there is a significant link between mobility and the ability to secure many of life's basic needs".[195]

Given that the underlying concept of merit goods is based on policy judgment, rather than an economic justification, developing a merit good "value" for low cost mobility is not feasible. Although substantive literature exists advocating transit as "meritorious" hence justifying their provision and public support, there is also a substantial body of evidence which supports the user-pay principle for transportation resources such that an efficient allocation of resources occurs. This rebuttal to the merit good argument assumes the perfect functioning of markets such that prices act as true signals of demand and supply. The limitation of the merit good argument for low-cost mobility, notwithstanding its validity from a policy perspective, is that it offers little in the way of analytic evidence to address this rebuttal.

Income Distribution and Low-Cost Mobility

The income distribution framework for considering the benefits of low cost mobility is based on the notion that the distribution of wealth that emerges from the market outcome is not considered by some to be socially equitable. Income distribution is most often facilitated through the tax system (income and wealth taxes, social security benefits etc.). It is generally recognized that the most efficient means of income distribution is through lump sum taxes and transfers which are least likely to create market distortions. However, it is also generally recognized that the most "efficient" instruments of income distribution are often not available to policy makers. This situation warrants a search for alternatives to meet the original objective. The efficiency/equity tradeoff best characterizes the income distribution framework for considering the benefits from low cost mobility.

This debate raises two issues. The first is whether the income distribution mechanism causes a reallocation of resources, compared to the market outcome. There is evidence in the transit literature that this is the case. In transit, individuals with low incomes are observed to have both a high income elasticity as well as a high price elasticity of demand.

Subsidized prices, therefore, may induce a demand response which requires additional resources (adding a route, buying more buses etc.). If there is a statutory requirement to maximize coverage, probably based on the merit good and income distribution arguments, then this may be a desired result.

The second issue is that studies show that the government tax and expenditure system has built in inefficiencies. What this means is that for every dollar spent it probably costs more than a dollar to collect that dollar in the first place, in terms of true resource costs. This is simply an indicator that spending of tax dollars on the direct provision of services may not be the most effective means of redistributing income.[196]

Much like the case of merit goods it is difficult to measure the value of low cost mobility benefits in the income distribution framework. With the exception of testing the provision of low-cost mobility as a income distribution mechanism (as compared to the tax system), there is little to indicate that it is a credible measure of value. However, unlike the merit good alternative, the degree to which low-cost mobility is used by individuals in lower income groups is observable. Therefore the notion of that transit is a mechanism of income distribution is testable.

Much of the basic mobility literature examines the relationship and/or interdependencies between the level of mobility and the quality of life or standard of living of individuals. Implicit in this type of analysis is a focus on the income distribution function of mobility. The overwhelming conclusion is that an individual's level of mobility greatly influences the quality of life they experience.

"Transport is seen as exerting a major influence on the quality of the lives of people and a certain minimum quality should be ensured".[197] Table 4.1 illustrates this point by displaying the per capita GNP of countries around the world and their respective per capita levels of mobility measured in terms of annual passenger miles for car, rail, air, and miles of railway. Within the top ten GNP ranked countries, North America places a higher relative importance on mobility as compared to wealthy nations in Europe and other parts of the world.

Bellah (1985) looked at individual socialization characteristics and found that the quality of life experienced in inner cities was undesirable in terms of such things as high crime rates, crowding and poverty.[198] However, moving to suburban areas was not possible, partially because of limitations in transportation systems to fulfill mobility needs. Mayer *et al* (1973) state that a "person's choice as to how to travel largely depends on whether he [or she] is healthy enough to walk, wealthy enough to run a car, or wise enough to live in an area with good public transport".[199]

A study conducted by Meyerhoff *et al* (1993) examined the mobility levels of the homeless and very poor in Los Angeles and found that, more often than not, the quality of life experienced is a direct consequence of the

Table 4.1 National Wealth and Mobility

Country	GNP per Capita	Travel Mobility	Travel Mobility Ranking	Freight Mobility
Switzerland	139	104	4	81
Sweden	119	96	6	151
USA	106	160	1	260
Netherlands	101	83	9	42
France	100	100	5	100
Canada	95	114	2	374
Australia	91	107	3	335
Japan	87	96	6	94
UK	63	78	10	47
Italy	53	86	8	49
Spain	43	54	11	44
Venezuela	31	24	13	36
Yugoslavia	24	32	12	55
Brazil	18	18	14	23
Mexico	15	14	15	42
Colombia	11	6	16	47
Nigeria	6	5	17	5
Egypt	5	5	17	13
Pakistan	2	3	20	10
China	2	3	20	16
India	2	5	17	26
Bangladesh	1	2	22	3

Source: Kenneth J. Button, *Transport Economics, 2nd Edition*, (Westmead: Saxon House, 1993), p. 19.

level of personal mobility. "[A]ccessibility to transportation often constrains one's ability to satisfy even the most rudimentary needs, such as food, shelter and medical care".[200] This coincides with Button's observations: low per capita GNP reflects low levels of mobility. The study found that more than 25 percent of the mobility problems were encountered when trying to access employment and that "nearly 40 percent of the employed respondents depended on public transit to provide access to work".[201]

Kain (1992) also observed employment accessibility difficulties. He examined the spatial mismatch hypothesis[202] of housing policies, and observed evidence revealing "low levels of auto ownership by minority residents in central cities is an important reason for low employment rates".[203] He discovered that many employers have been moving away from the Central Business District (CBD) into the suburbs. Transportation infrastructure is generally designed to bring labor into the CBD rather than the reverse. Without reliable outbound transportation, these jobs become inaccessible for minorities and/or low income households. One solution suggested by Hughes (1989) calls for the restructuring of transportation systems facilitating outbound journeys to work. In some cases, employers cooperating with transit authorities are paying for much or all of the cost of new routes.[204]

Profile of Transit Users

Figure 4.1, a profile of transit use in the United States, provides an indication of transit's role as a mechanism of income distribution. Although urban transit's share of total trip-making appears small at first blush (two percent of daily trips in 1990[205]), it provides more than seven billion journeys annually and facilitates vital activities of daily living including family life, work, religious activities and getting to and from places of education.

The 1990-91 NPTS Survey shows that that transit riders have much lower incomes on average than do auto users or households in general. Whereas 28 percent of transit riders have an annual income that is lower than $20,000, only 16 percent of private owned vehicle users have incomes this low (See Table 4.2). This changed little from the 1977-78 NPTS which showed that transit matters to low income and other disadvantaged people. For example, buses and subways handle fully 42 percent of all trips made daily by people with disabilities who lack regular access to an automobile.[206] Urban transit users earn, on average, well beneath the median household income of the urban public at-large. Compared to median household income in 1993 of $31,241, the median household income of transit users in that year was $25,900. Users of fixed- route bus service come from households whose median income stood at $24,960 in 1993–20 percent beneath the urban median. At $28,100, the median household income of rail transit passengers is eight percent higher than that of bus riders but still only 90 percent of the urban household median. Fixed-route bus transportation serves a disproportionately large share of the

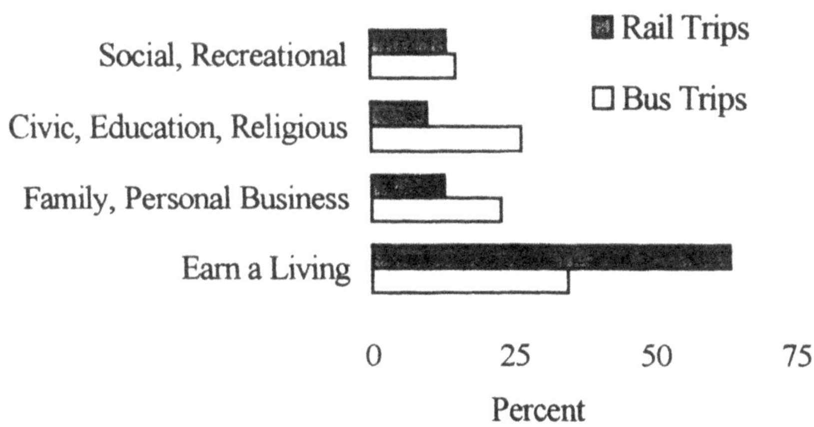

Figure 4.1 Transit Purposes by Mode

Source: U.S. Department of Transportation, Nationwide Personal Transportation Survey, 1990.

urban poor for whom access to an automobile is limited. In 1990, riders with income less than $10,000 accounted for 13 percent of bus and streetcar users, compared to 4 percent of subway users, and 3 percent of commuter rail users. Bus riders were found to have by far the lowest average incomes of any modal user group, and commuter rail riders by far the highest average income.[207] The 1990-91 NPTS also finds that bus riders have the lowest incomes of any modal users, with 30 percent of them having incomes less than $20,000, and only 3 percent of them with incomes of $80,000 or more (see table 4.2).

Transit is especially important to people with multiple socio-economic disadvantages. Individuals with disabilities and no access to a car—those making 40 percent of their daily journeys by transit (see below page 188)–live in households whose median household income lies beneath $12,000 annually (in 1990 dollars).[208]

The availability of a car appears, in Table 4.3, to affect transit rail patronage as much as bus patronage. Riders without autos account for about 37 percent of rail passengers and 40 percent of bus patronage. However, Table 4.4 indicates that few bus passengers are people who can

Table 4.2 Composition of Mode Users by Household Income, 1990

	Annual Household Income (Thousands)							
	less than $10	$10 –20	$20 – 30	$30--50	$50- 80	$80 & over	N/A	Total
POV	5	11	14	25	17	6	21	100
Transit (total)	12	16	14	16	10	4	28	100
Bus & Streetcar	13	17	13	16	9	3	28	100
Subway	4	12	14	20	17	6	28	100
Commuter Rail	3	5	17	16	18	13	29	100

Source: 1990 Nationwide Personal Transportation Survey.

Table 4.3 Transit Trips by Mode and Auto Availability, 1990

	Annual Transit Trips (Millions)		
	Bus	Rail	Total
Total	3,543.2	1,349.3	4,892.4
Percent of Total Trips	72	28	100
People With No Auto Available	1,411.1	492.5	1,903.6
Percent of Mode	40	37	39
People With Auto Available	2,132.1	856.7	2,988.8
Percent of Mode	60	63	61
Total	100%	100%	100%

Source: 1990 Nationwide Personal Transportation Survey.

easily afford cars of their own. Of 27 passengers without cars who ride buses, 20 have income near or below poverty. (i.e., $20,000 in 1990). In the higher income groups, people with cars add more bus patronage than people without cars, presumably on an incidental basis, unrelated to auto availability per se.

The Economic Value of Low Cost Mobility

The individual's travel objectives are obtained only at a price, which includes the direct money cost people pay plus the cost to them of using up time and of physical effort and inconvenience. The economic value people obtain from mobility is the value they derive from their journey purposes (from work, shopping, going to school etc.), not from the journeys themselves. This value always exceeds the journey price; if it were not so the trips would not be made.[209] The net value people obtain from mobility is equivalent to the derived value as defined above minus the journey price they pay.

Table 4.4 Transit Bus Trips by Auto Availability and Household Income, 1990

	Annual Household Income							
Bus Trips (Million)	Under $10	$10-20	$20-30	$30-40	$50-80	Over $80	N/A	Total
Total	545.01	658.56	465.61	503.68	244.14	92.69	966.73	3,476
Auto Not Available	389.85	304.83	119.41	112.21	20.37	5.76	427.51	1,380
Cumulative Percent	11%	20%	23%	27%	27%	27%	40%	40%
Auto Not Available	155.16	353.73	346.20	391.47	223.77	86.92	539.22	2,096
Cumulative Percent	4%	15%	25%	36%	42%	45%	60%	60%

Source: Nationwide Personal Transportation Survey, 1990.

The economic value people obtain from low cost mobility is the value they derive from their journey purposes (work, shopping, going to school etc.) minus the fares they pay. In 1993, the economic value of transit as a source of low cost mobility was $33.7 billion.

The net economic return to the nation from its investment in low cost mobility was an estimated $17.5 billion in 1993 ($33.7 billion in the value obtained by individuals minus the $16.24 billion in nation-wide public expenditures to operate, maintain and invest in transit systems).

Low-income or disadvantaged individuals value their mobility as it directly affects their standard of living. Economic theory suggests that these individuals make decisions to increase their quality of life. As Meyerhoff et al. (1993) discovered "most people deemed transportation so critical to their own needs that they were willing to allocate a portion of their already extremely limited budget toward fulfilling their travel needs".[210]

One in every ten people in Los Angeles county depend on some form of public assistance. Meyerhoff et al. (1993) surveyed 203 public assistance recipients finding that the average monthly income was $360, the average rent was $158 and the average expenditure on transit was $35 (roughly 10 percent of their budget). That left $123 for food, utilities, clothing and other personal items.[211] In order to maintain some form of public assistance, certain tasks had to be completed by the recipient. Some of these tasks which include performing 17 days of assigned county work per month and completing 24 job searches in an 8 week period, obviously require a flexible level of mobility. In the most extreme cases the study revealed that individuals were forced to forego food and shelter in order to pay for transportation.

Foregoing the most basic needs to increase an individual's mobility surely reflects the high priority individuals give to mobility. Dittmar and Chen (1995) state that "transport is not the effect of poverty, but rather a root cause of poverty".[212] Access to mobility becomes not only important, but necessary in securing an acceptable standard of living.

According to Mayer *et al* (1973) levels of quality and cost limit the use of public transportation, especially to those groups most dependent upon its services. The young, the elderly and the poor become public transportation's 'captive market'.[213] It is this captive market that also seems to pay the greatest fares relative to marginal costs of services they use. They are people who travel more often at off-peak times, take irregular route journeys i.e. outward CBD bound rather than inward to CBD, patronize lower cost bus service, and take shorter trips.

However, evidence shows that peak period, and longer journeys generate higher operational costs.[214] Generally, such journeys are completed by higher income riders making journeys from the suburbs into the CBD. When fares are undifferentiated between peak and off-peak times, the resulting phenomenon is that low income and minority riders are cross subsidizing higher income and non-minority riders.[215]

Almanza and Alvarez (1995) examined the impacts of transportation facilities situated in low-income communities and minority communities. They discovered that these facilities are frequently designed to expedite the transportation needs of rush hour riders, failing to address the transportation needs of the disadvantaged or those directly surrounding the facilities. They conclude that low-income people and minorities are usually left out of the transportation planning process, and have little or no control over the environmental and economic impacts they suffer as a result of placement of those facilities in their communities.[216]

Such circumstances only prove to further the delineation between social classes and income levels. As Roseland (1992) states:

> "The sustainable city is one that achieves a steady improvement in social equity, diversity and opportunity and 'quality of life,' broadly defined. Economic, fiscal and sectoral policies, however, often have the unintended effects of reducing all of these and increasing social polarization and cultural and economic barriers between groups".[217]

As is more thoroughly examined in later sections of this paper, fiscal policy designed to reduce spending in the transportation sector often results in increased spending in other economic sectors, such as welfare and unemployment. Dittmar and Chen (1995) found that transit has received approximately $50 billion in expenditures since 1965.[218] Mobility vulnerable programs such as Medicaid, unemployment and social services received $443 billion in 1994 alone.[219]

This literature strongly suggests that the value to society of affordable transit mobility exceeds the revenue these service produce in the farebox. How much the value exceeds fares is unknown. However, economic theory provides a number of tools for calculating benefits from subsidized economic activities.

Estimating the Economic Value of Low Cost Mobility

The next sub-section of the chapter presents estimates of the economic value of "low cost" mobility to people with low income, lack of ready access to automobiles, physical disabilities and other barriers to their ability to travel in urban areas. The focus is mass transit, with special reference to urban fixed-route bus transportation. The term "low cost" refers to the fare people pay, the affordability of travel. The term "low cost mobility" can be re-phrased "affordable mobility" without loss of clarity, a useful synonym in sentences that would otherwise refer simultaneously to supply costs (labor, capital) and "low cost" as a connotation of an affordable fare.

The full costs of making affordable mobility available are not reflected in the price people pay; indeed, public financial support is a key mechanism through which transit is employed as an instrument of affordable mobility. But they are costs all the same. They include the capital expense of purchasing vehicles and constructing infrastructure; the operating expense of paying drivers and managers and fueling and insuring vehicles; and the maintenance expense of keeping vehicles and facilities in repair. If these costs, in their entirety, lie beneath the net value people obtain from mobility as defined above, the economy can be said to be earning a good return from its low cost mobility investment. If the full costs of supplying affordable mobility exceed the value people derive from it, the nation's cities taken together would be on a path to lower growth and living standards. The next sub-section concludes with this cost-value comparison.

The Role of Personal Expenditure on Transit in Household Budgets

The information presented above relates to trip-making. Information about household *expenditures* on transportation reveals additional insight into the role of transit in the lives of poor people. First, mobility makes a sharply disproportionate claim on the household budgetary resources of the poor. This is clear from the Consumer Expenditure Survey statistics presented in Figure 4.2, Figure 4.3, and Figure 4.4 which show motorized transport costs low income households a *larger* share of earnings than higher income households. This is true of all modes including autos, transit and even taxis. Even though poor people make fewer daily auto trips than those with high earnings,[220] they spend well over 50 percent more of their available budget on getting around by car than do people from higher income households. The importance of bus transportation to poorer people is strikingly evident in the household expenditure data presented in Table 4.5. As household incomes rise from the lowest levels to about $15,000 (in 1994 dollars) spending by household members on bus transportation rises disproportionately. Thereafter expenditure on bus transportation falls and continues on a downward trend as household incomes continue to grow. This pattern of expenditure reflects a propensity to travel by car as household income rises.

The cause-and-effect dynamic underlying this pattern is doubtless "two-directional" and mutually reinforcing with (i) rising income creating more opportunity for the poor to participate in life activities *and* (ii) more income-earning opportunities for the poor, as they arise, creating greater travel This very high "income elasticity" of demand for transit among low income households indicates that the poor forgo a great many life activities they value highly and that bus transportation is a critical outlet for such activity as it becomes increasingly affordable.

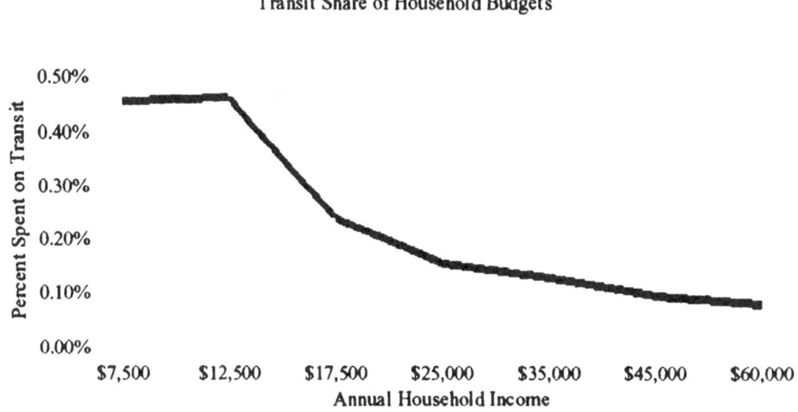

Figure 4.2 Average Household Urban Transit Expense

Source: Bureau of Labor Statistics, *Consumer Expenditure Survey*, 1994.

Role of Taxis as a Means of Low Cost Mobility

Taxis served 422 million passengers in 1990,[221] increasing the total number of public transportation journeys made in that year by 8.6 percent. Though not widely regarded as a "public" mode of transportation or as a source of low cost mobility, taxis actually serve both roles, *de facto*, due to the way poor and other disadvantaged urban dwellers use them. Figure 4.3 indicates that the poor spend substantially more of their income on taxis than people from higher income households. In fact, they spend more in absolute terms as well and thus account for a disproportionate share of the nearly half a billion taxi journeys made annually.

The pattern of taxi use outlined above stems from the lower average rate of auto availability among poorer households and the unavailability of bus transportation for certain kinds of trips (See Table 4.7). Inner-city residents with domestic employment in the suburbs, for example, often find that bus-routes geared to downtown centers do not serve their destinations, thereby compelling them to choose taxis for at least part of their journey to work.[222]

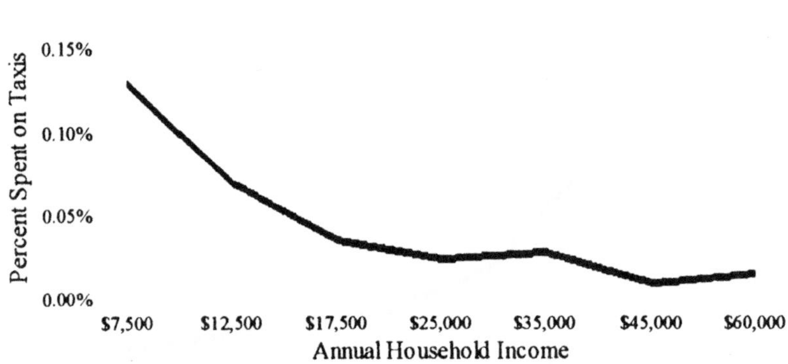

Figure 4.3 Average Household Taxi Expense

Source: Bureau of Labor Statistics, *Consumer Expenditure Survey,* 1994.

Low income people share something in common with other disadvantaged groups in their reliance on taxis. People with disabilities use taxis for three percent of their daily trips at an average fare of $9.95 (in 1994) despite their low incomes. This rate of taxi use is higher by a factor of 15 than the rate of taxi use by people without disabilities (who use taxis for only 0.2 percent of their travel requirements).[223] Among the disabled, taxis are especially important to wheelchair users and people with visual impairments. Each of these groups use cabs for seven percent of their daily trips (35 times the general public's consumption of taxi travel).[224]

Predictably, wheelchair users are the most frequent users of "ambulette" services. These are wheelchair-accessible commercial taxi services providing door-*through*-door assistance with specially-trained drivers and specially-equipped vehicles. Vehicles are equipped for people who travel in wheelchairs or on gurneys and who carry oxygen, dialysis and other life-support equipment at all times. When neither "ADA" dial-a-ride[225] nor accessible bus service is available or appropriate to their needs, people will turn to commercial services at fares that exceed $45 per one-way trip.

While many of these journeys have a medical purpose, some are made to get to and from sports, social and entertainment events.[226] Young gurney users on life-support devices attending rock concerts, amusement parks, baseball games and so on will frequently travel to and from such events by this means and at their own cost.

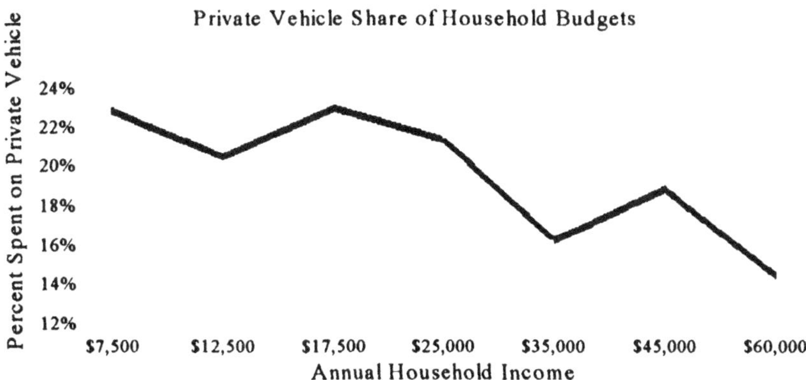

Figure 4.4 Average Household Auto Expenses

Source: Bureau of Labor Statistics, *Consumer Expenditure Survey,* 1994.

As indicated in the first section, the value that people obtain from mobility is the value they derive from their journey purposes. From this we need to subtract the journey price they pay in order to measure the net economic benefit transit passengers obtain from low cost mobility. Conceptually, a good index of net benefit is the maximum amount individuals are actually willing to pay to make journeys, minus the fare actually charged.

Willingness to Pay as an Expression of Value

In reality, different people are willing to pay different amounts, even for identical journey purposes. Everyone faces life with different social and economic circumstances, different physical attributes and mental abilities and different interests, attitudes, beliefs and convictions. It is both circumstance and preference that enters into an individual's personal calculus of "willingness to pay" as an expression of the value placed on each of the myriad of life's everyday activities which occasion travel.

Table 4.5 Average Annual Expenditures on Mass Transit by Income and as a Percent of Income, 1994

Household Income	Median Household Income	Average Annual Expenditure on Mass Transit	Percent of Income
$5,000-$9,999	$7,500	$34.30	0.5%
$10,000-$14,999	$12,500	$57.71	0.5%
$15,000-$19,999	$17,500	$41.68	0.2%
$20,000-$29,999	$25,000	$38.71	0.2%
$30,000-$39,999	$35,000	$44.84	0.1%
$40,000-$49,999	$45,000	$43.16	0.1%
$50,000-$69,999	$60,000	$48.25	0.1%

Source: Bureau of Labor Statistics, Consumer Expenditure Survey, 1994.

Table 4.6 Propensity to Use Taxi Service by Income

Household Income Yr ($Thousands)	7.5	12.5	17.5	25	35	45	60
Elasticity	1.71	0.61	-0.57	-0.57	0.00	0.00	0.00

Source: Hickling Lewis Brod Economics Inc. Estimates from data supplied by Bureau of Labor Statistics, *Consumer Expenditure Survey*, 1994.

People without cars who demonstrate a willingness to pay $45 or more for ambulette service are proportionately few in absolute numbers but they would value the availability of affordable transit (a $1.50 bus ride, for example) more than those choosing to pay $9.95 for a cab ride when affordable transit is unavailable. The latter group, a sizable number of people as indicated earlier, will value the availability of transit more than those for whom a $1.50 average bus fare is only marginally lower than the maximum amount they are willing to pay for public transportation. This group is larger-still, a conclusion that stems from the fact that bus fare revenues per linked trip have in actuality tended to settle at the $1.50 level,[227] reflecting local policies to stabilize a long-term declining trend in bus ridership.

Figure 4.5 illustrates that it is the collective, or aggregate valuation of affordable transit by all such diverse individuals that represents the economic value of low cost mobility.[228]

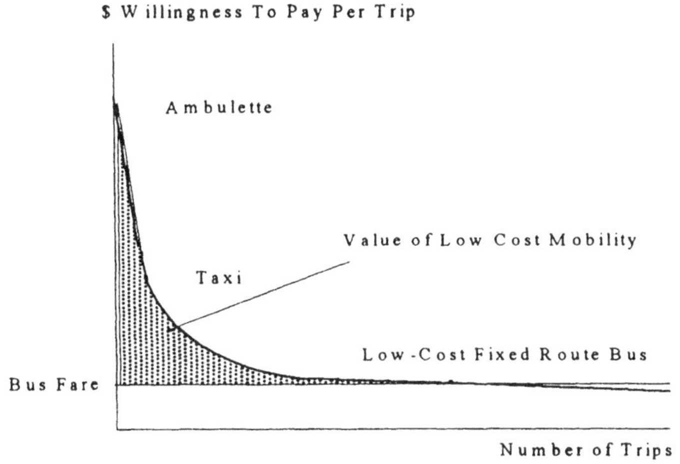

Figure 4.5 The Demand–Value Curve for Transit

Source: Hickling Lewis Brod Economics, Inc., 1996.

Market Data

The starting point in estimating the aggregate value of transit as low cost mobility is transit market data. Transit, (bus, subway and commuter rail) met the demand for 7.2 billion one-way trips in 1993 at an average fare of just under $1.50.[229] Fare increases around the country over the past several years have revealed a market fare elasticity of about -0.3 which means that each one percent increase in fare results in a 0.3 percent reduction in ridership. The -0.3 elasticity is an *average*. Some passengers are willing to pay a great deal more than $1.50 and fare increases have no impact on their use of transit. Others will already be at the maximum they are willing to pay and the next fare increase leads them to stop using transit altogether.

An important insight about the elasticity of demand among low income transit users springs from the analysis of Consumer Expenditure Survey data presented above. As shown in Table 4.5, high income earners tend to reduce their spending on transit as their income continues to improve, while those with low incomes use extra earnings to *increase* their spending on transit. The income elasticity of 1.7 means that the poorest passengers increase their transit use in amounts that actually exceed, proportionately, the additional earnings. This divergence in the way transit users of different incomes respond to changes in their monetary circumstances implies similar differences in the way they respond to changes in fare. A change in fare will provoke a relatively greater response among passengers from poorer households compared to those with higher incomes.

Estimated Value of Low Cost Mobility

The results are shown in Figure 4.6(a) and (b). The analysis indicates that in 1993 the value of transit as a mode of low cost mobility was $33.7 billion, this being the difference between the expenditure on bus travel people were willing to incur and the amount they actually incurred at the average fares charged in that year.

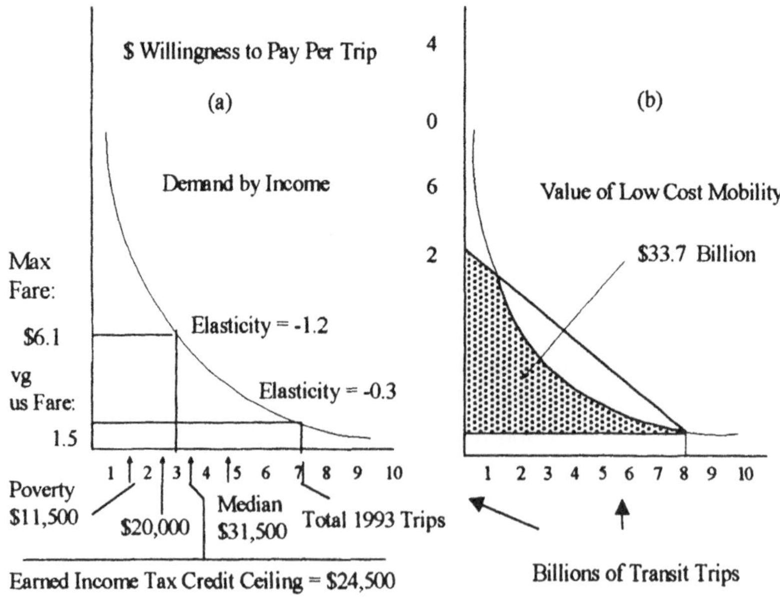

Figure 4.6 Income and the Economic Value of Low Cost Mobility

Source: Hickling Lewis Brod Economics, Inc., 1996.

As seen in Figure 4.6(a), the estimated willingness to pay rises to levels close to $50 per trip for a small number of passengers and a valuation of low cost mobility based on this curve is substantially greater than $33.7 billion. While there is indeed evidence of such demands among ambulette users (see above), the practical reality is that such trips cannot be accommodated on fixed-route transit buses even where vehicles are wheelchair-accessible and routes and frequencies otherwise appropriate. Figure 4.6(b) indicates that a simple linear interpretation of the curve suggests that the maximum willingness to pay for low cost mobility trips by those for whom buses are a feasible design option lies at about the $11.50 per trip level. This magnitude is consistent with the fact that many trips for which bus service is unavailable (but otherwise feasible) are made instead by taxi at average fares of around $10.00.

Although transit users are, on average, financially less well off than the general public, the most severely disadvantaged represent a sub-set of all bus and rail passengers for whom transit's total mobility value is an estimated $33.7 billion. Drawing distinctions however is inevitably somewhat arbitrary. Among the 30 percent of all transit trips made by people in households earning less than $20,000 a year (about 2.89 billion trips in 1993) the estimated maximum willingness to pay for a low cost mobility trip is $6.1 (Figure 4.6(a)). The total value of low cost mobility to this sub-group is $20.5 billion as shown in Table 4.7.

Ability to Pay and the Estimate of Value

The valuation above hinges on the validity of peoples' own assessment of what is desirable from their point of view. For low income people, such assessments are inevitably conditioned by their constrained economic circumstances and it is well known that poverty can lead people to make personal choices that are not the same as those society might make for them in order to maximize their contribution to economic growth and social vitality (see section on Merit Goods). This indeed is one view of why governments will elect to subsidize certain commodities, such as food, namely to induce people to consume more than they otherwise might and thus advance the community toward a social optimum.

Whether the amount of travel that low income people and other disadvantaged groups elect to engage in is enough, given their circumstances, to maximize their contribution to economic growth and vitality cannot be ascertained in any scientific way. What *can* be ascertained however is whether the value of travel that is induced by public financial support for low cost mobility exceeds the costs of such support. If so, it can be said that a positive contribution to economic growth and vitality is being achieved. This question is addressed next.

Table 4.7 Consumer Surplus for Transit Users, by Income Group

	Income Group Ceilings					
	Poverty Line $11,500	$20,000	EIC* $24,500	Median HH Income $31,500	Median Urban HH Income $34,500	All Riders
Trips (Billions)	1.54	2.89	3.53	4.47	4.80	7.20
Elasticity	-1.8	-1.2	-1.0	-0.8	-0.7	-0.3
Consumer Surplus (Billions)	$13.36	$20.49	$22.63	$26.11	$27.18	$33.70
Cumulative Consumer Surplus	40%	61%	67%	77%	81%	100%

*EIC: Earned Income Tax Credit to delineate households that are recipients of favorable National tax policies owing to their low income status.

Source: Hickling Lewis Brod Economics estimates.

Economic Return From Current Investment in Low Cost Mobility

The $33.7 billion in low cost mobility benefits is obtained at a cost, namely the operating and capital costs incurred in the provision of transit services. In 1993 the nation spent a total of $15.97 billion to operate and maintain transit services (including capital depreciation allowances) and laid out lump sums totaling $5.35 billion for new capital equipment and facilities. Of course, new equipment and facilities provide service for more than a year. Assuming a 20 year life for these investments yields a $268 million annualized amount; this added to the $15.97 billion indicates total transit expenditures in 1993 of $16.24 billion. The net economic return to the nation from its mass transit investment in low cost mobility is thus positive at an estimated \$17.5 billion annually (\$33.7 billion minus \$16.24 billion, at 1993 service levels and in constant 1993 dollars).

As discussed in Chapter 2, however, the avoidable costs of transit generated by its low cost mobility role are substantially less than ithe industry's total annual costs. Chapter 2 estimated the costs of aggregate United States mobility costs as \$6.4 billion in 1993. Furthermore, transit subsidies to middle class Americans legitimately belong in benefit categories other than low cost mobility. Accordingly, the consumer surplus of \$22.6 billion to transit passengers earning less than the Earned Income Tax Credit ceiling (from Table 4.7) reflects national public policy in the United States to bolster the incomes of the working poor.

Basis for Formulating the Transit Demand Curve

As transit fares rise and the money cost of travel increases in importance relative to the time and effort components of travel cost, the theory of generalized cost predicts that the market fare elasticity will rise accordingly. Simply stated, when fares are already "high", a one percent increase will precipitate a larger proportional effect on demand than a one percent increase when fares are "low".

$$\eta = \frac{dT}{df}\frac{f}{T} = a + bf \qquad \text{eq. 1}$$

In words, the elasticity (denoted by the Greek letter eta) of trips (T) with respect to fare (f) is a function of fare.

There are strong empirical as well as theoretical foundations for the expectation that the marginal impact of fares on demand increases as fare levels rise. As shown in the second section of this chapter, people from low income households increase their use of transit when their incomes rise by a much larger amount (proportionately) than higher income people. It is well known that the marginal utility of an extra dollar is much higher for the poor. One can take the evidence regarding income elasticity as empirical confirmation that low income transit users are more responsive than higher income people to any transit-related change in their financial circumstances, including change induced by fare increases or reductions. The same finding has been reported in other studies.[230] The differential eq.1 implies the general demand function,

$$\ln T = k + a \ln f \text{ - } bf \qquad \text{eq. 2}$$

a special case of which is:

$$lnT = k \text{ - } bf \qquad \text{eq. 3}$$

Eq. 3 implies that fare elasticity is directly proportional (inversely) to fare level, that is, dT/df (f/t) = bf. Eq. 2 is more general than eq. 3, indicating that fare elasticity *may* in fact be indirectly proportional to fare level and it is in this sense that eq. 3 is a special case of eq. 2. Since the empirical data available are too limited to test the more complex possibilities of eq. 2 the analysis here adopts the assumption of proportionality between fare elasticity and fare level given by eq. 3. The approach to sensitivity analysis related to these assumptions is discussed in Annex 4.1.

Estimating the Demand Function

Given the current demand for transit, current fare level and the current fare elasticity, eq. 3 will give the estimated market demand curve for transit. In 1993 transit served 7.2 billion trips at an average one-way fare nationally of just under $1.50. The industry-average fare elasticity is -0.3. This gives the following result for eq. 3, f = (23.0–LnT)/0.2. The equation is written in inverse form corresponding to the conventional demand curve, shown for this enumeration of eq. 3.

An important check on the likely validity of the function shown in Figure 4.6(a)) are its implications about consumer behavior over a wide range of circumstances. An estimated 2.89 billion trips in 1993 were made by people from households earning under $20,000.00 annually. As shown in Figure 4.6(a), the demand function implies that a fare level of $6.1 would clear the transit market to this volume of trips and that the fare elasticity would be -1.2. This result conforms to the empirical evidence presented in the second section of this chapter. Low income people, including those without the use of a car, elderly people and people with disabilities, turn disproportionately to taxis, at average cab fares in excess of $9.00, when transit is unavailable to them. Their "willingness to pay" fares of $6.1 or more is thus confirmed. It is also reasonable to expect that, once fares rise to the $6.00 level, the fare elasticity will rise above -1.0, implying that further increases would lead to revenue losses. The fare elasticity of -1.2 implied by the demand function at this fare level is also consistent with the fact that the estimated income elasticity of transit among low income households is very high.

Estimating the Value of Low Cost Mobility in 1994

The difference between willingness to pay and fares actually paid by transit passengers represents the economic value of transit. Since transit users are typically from low income households this value is, in principle, equivalent to the area under the demand curve and above the $1.50 fare level in Figure 4.6(a).

Concerns with this calculation arise at very high fare levels, however. A mathematical characteristic of eq. 3 is that it rises and falls asymptotically to the axes, which means that some demand would be predicted at any fare level. As it happens, the function approaches the fare axis at around the $30.00 per trip mark and the function predicts trivial demands at fares above this level. Even so, while evidence in the main text indicates that some low income and other disadvantaged people are willing to pay such amounts, such trips typically involve specially-equipped vehicles and specially-trained drivers that conventional transit cannot be expected to provide, even under the Americans with Disabilities Act (vehicles for gurney users on personal business, such as attending sports and entertainment events).

To guard against overstating the value of conventional transit, the analysis thus modifies the demand function in Figure 4.6(a) and estimates a "kink" in the curve above the $6.1 level. The $6.1 level was chosen since the point on the curve denoted by "$6.1 fare; 2.89 billion trips;-1.2 elasticity" has been validated above. The demand curve above the $6.1 fare level is estimated as a straight-line extrapolation between 7.2 billion trips and 2.89 billion trips, as shown in Figure 4.6(b).

The demand function shown in Figure 4.6(b) is the basis on which we estimate the economic value of transit as a mode of low cost mobility. The total consumer surplus in 1993, the shaded area in Figure 4.6(b) is equivalent to $33.7 billion.

Sensitivity of Results to Demand Assumptions

Sensitivity analysis is conducted to account for the uncertainty surrounding key parameters, namely, the elasticity with respect to fare and the corresponding willingness to pay. The sensitivity analysis is conducted by identifying feasible pairs of both elasticity and fare which are within the bounds of empirical evidence and which do not violate the general form of the demand curve.

A series of tables in Annex 4.1 identify the feasible pairs of elasticity and willingness to pay for all transit users (using $20,000 as the low-income anchor point), and separately for each income category of interest.

The sensitivity analysis indicates a lower bound consumer surplus estimate is approximately $21 billion dollars. This point is the minimum consumer surplus estimate without the assumption that low income users are more responsive to changes in price. The high bound of the sensitivity analysis indicates a consumer surplus of approximately $64 billion. The estimate of consumer surplus under the original assumptions lies approximately in the middle of the range at $34 billion.

Cross-Sector Fiscal Impacts of Low Cost Mobility

The economic benefits of affordable mobility as presented above are "market" benefits in the sense that they accrue to the immediate users of transit. "Club" and "spillover" benefits also arise from transit-based affordable mobility. These are manifest principally in the form benefits to non-transportation social service programs. Studies by Carr, et al. (1993) and others have shown that transit can relieve demand and financial pressure on non-transportation social safety net programs. The reverse is also true; cuts in transit budgets lead either to increased expenditures on non-transportation social service expenditures (health, nutrition and unemployment support programs) or, alternatively, to reduced benefits for those in need of such programs. Our analysis, presented below, finds that major cut-backs in transit create huge financial pressures and dislocation in such programs. Information and other barriers are such that governments rarely respond to such "cross-sectoral" effects with counter-balancing budgary increases in the dislocated sector. This means that transit cuts harm not only transit recipients of affordable mobility, but users of basic health, nutrition and unemployment safety-net services as well.

Previous Empirical Studies of Cross Sector Fiscal Impacts

This section focuses on two studies examining the existence and prevalence of cross-sector benefits generated by low cost mobility investment in the United Kingdom; completed by the MVA Consultancy (MVA) and The Centre for Logistics and Transportation Cranfield University (CCLT). The section also briefly examines examples found in the United States.

Cross-sector benefits are defined to be "economies achievable in another sector of the economy as a result of expenditure in the transport sector".[231] The two British studies focused on the impacts in the delay of moves made by the elderly and/or disabled, from in-home services to institutional care and substitution of improved mobility for in-home services resulting from the provision of accessible public transport (APT).

Both studies conclude that significant cross-sector benefits arise of between $45,048 and $60,064 per 1,000 persons resulting from "the relationship between mobility and the use of in-home services".[232] Both studies also suggest that these estimates were under rather than over-estimated. The researchers discovered considerable latent demand in APT services to be the cause of an underestimation of cross-sector benefits. Many individuals interviewed either were not aware of APT service existence or experienced availability restrictions. However, this suggests "that cross-sector benefits would be proportionately higher if APT services were targeted and more widely available".[233]

The MVA Consultancy conducted three surveys in North Birmingham and Coventry. They surveyed people living in institutional care; both users and non-users of APT services and of informal carriers--completed from December 1990 to September 1991. As before, cross-sector benefits are defined as savings in one sector as a result of expenditure in another. The MVA hypothesis was that transportation investment results in savings in health and social service provisions if:

transition from in-home services to institutional care is delayed, and

substitution of APT services for in-home services exists.

Examining the cost differentials between institutional and in-home provisions illustrates the existence of cross sector benefits. "[T]he costs of in-home care range from 25 percent to 75 percent of those of institutional care".[234] Any transition delay between these care provisions results in benefits. The surveys indicate, however, that variables influencing the decision to move from in-home care to institutional care are very complex and often are not mobility dependent. The level of personal mobility may be a contributing factor but "the provision of APT services does not or would not affect either the decision to move, or the timing" thereof.[235]

The research indicates a strong relationship between the ability to use public transport and the use of in-home services. In general, the greater the difficulty in using public transport experienced, the greater the need and/or use of in-home services. Figure 4.7 presents these results graphically. However, APT services are more accessible to the disabled and therefore may decrease the percentage of individuals needing in-home services. The researchers found that if APT services were not provided 16 percent of current riders would require in-home services.[236]

The substitution effects of using transport services rather than in-home care were estimated by MVA as follows: in North Birmingham, per 1000 disabled individuals, annual cross-sector benefits range from $6,781 to $70,236; in Coventry–$7,008 to $70,075.[237] Population of disabled individuals residing in these cities are 31,000 and 15,000 respectively.

In a similar study, The Cranfield Centre for Logistics and Transportation (CCLT) conducted two studies in Strathclyde, South Yorkshire and London from 1985 to 1989 to examine cross-sector fiscal impacts by observing the use of APT services. They differed only in the size of the survey population with the second study being the larger of the two.

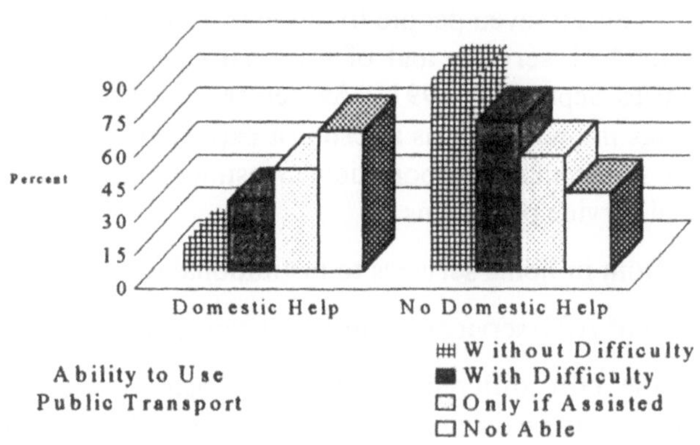

Figure 4.7 Mobility and the Need for Domestic Help

Source: Carr, et al, 1993, p. 15.

Similar to the MVA study results, CCLT found a significant relationship between the ability to travel and the number of in-home visits received. "[A]t the aggregate level for every two additional activities which a person was able to get out to, there was a reduction of about one in-home visit".[238] In-home visits cost approximately twice as much as facility visits (see Table 4.9).

In order to quantify cross-sector benefits, the Cranfield Centre examined specific types of in-home services. In so doing, they determined the average number of yearly visits by each service type as well as the costs of such visits. Then, to determine the possible number of visits replaced by transport use, each in-home service was rated as high, high/medium, medium and low with reference to their degree of possible in-home visit reduction.[239] Table 4.8 summarizes the research findings. By applying specified hourly costs and duration of visits to the number of reduced visits, CCLT quantified cross-sector benefits resulting from reduced spending in health and social service sectors (Table 4.9).

Table 4.8 Potential Reduction of Home Visits

Category	Service	Percent Reduction in Home Visits	Equivalent No. of Home Visits Saved per 1,000 People
High	Optician	60	12
	Podiatrist	60	528
	Hairdresser	60	942
	Library	60	504
	Meals on wheels	60	3,348
High/Medium	Occup. Therapist	50	7
	Physiotherapist	50	12
Medium	Doctor	35	1,162
Low	Home help	10	2,854
	Health Visitor	20	614
	Social Worker	20	55
	Total		10,038

Source: Carr, Melanie et al. "Cross-Sector Benefits of Accessible Public Transport", Environment Resource Centre, Transport Research Laboratory, Crowthorne, Berkshire 1993, p. 47.

CCLT was conservative in estimating the cost savings associated with substituting transit for home visits. For example, even though fewer home visits by mobile library services, such as "bookmobiles", would be required at the margin, CCLT assumed negligible savings. In the case of meals-on-wheels, CCLT recognized the role of volunteer labor and donated food in the delivery of such services and reduced the potential cost savings of transit substitution accordingly.

Table 4.9 Cross-Sector Benefits of Reduction in In-Home Visits

	Domiciliary Visits				Visits to Facilities			Net Annual Saving
$U.S.	Hourly Cost	Home Visits Replaced	Minutes per Visit	Total Cost	No. of Visits	Minutes per Visit	Total Cost	Per 1,000 People
Optician	18.92	12	40	152	12	22	83	69
Podiatrist	18.92	528	40	6,660	528	22	3,662	2,997
Hair-dresser	6.01	942	90	8,487	942	60	5,658	2,829
Library	N/A	504	N/A	526	504			526
Meals on Wheels	N/A	3,348		1,760	3,348		N/A	1,760
Occup. Therapist	18.92	7	60	132	7	22	48	84
Physio-therapist	18.92	12	60	227	12	22	83	144
Doctor	49.85	1,162	30	28,964	1,162	16	15,448	13,516
Home Help	10.06	2,854	60	28,714				28,714
Health Visitor	16.52	614	60	10,142	614	35	5,916	4,226
Social Worker	16.52	55	60	908	55	35	530	378
Totals		10,038		86,671	7,184		31,428	55,242

Source: Carr, Melanie, et al, "Cross-sector Benefits of Accessible Public Transport", Environment Resource Centre Transport Research Laboratory, Crowthorne, Berkshire 1993. p. 48.

In summary, cross-sector benefits of $55,242 per 1000 people per year are realized when 10,038 home visits are replaced by low cost mobility. Facility visits also result in economies that are difficult to quantify in that they are somewhat intangible. For instance, it is intuitive that surgeries performed in-hospital will be of greater quality then those performed at home due to higher accessibility to premium quality equipment and possibly greater assistant staff availability. Facility visits are of shorter duration, facilitating a greater number of patients to be seen, resulting in a reduction per person and/or visitor in the cost of providing such services. CCLT asserts "if individuals are able to conduct their own personal business, shopping and social activities, their calls on in home services will be significantly less, particularly on home help, podiatry, health visitors and doctors".[240]

United States Experience

The General Accounting Office (GAO) completed a report detailing State efforts to expand home services while limiting respective costs. They examined the efforts of the nation and the states of Oregon, Washington and Wisconsin. In the nation, expenditure for institutional care facilities greatly outweighs the amount consumed by home and community-based care.[241] Figure 4.8 provides a comparison of federal Medicaid expenditures for institutional and home care. The GAO (1994) also reports that by restricting capacity growth in institutional facilities and expanding home and community based services, more people were able to be served than possible if the reverse had occurred. Institutional care is far more expensive per person than are home or community based provisions. Table 4.10 provides evidence on the cost differences of these service programs.

Both the MVA and CCLT studies indicate that if transportation services are substituted for or could delay the transition to institutionalized facilities, cross-sector benefits would be realized. Although they were unable to directly quantify such benefits, there was no doubt that they do exist. The evidence collected for the United States suggests a similar potential for benefits.

Other Areas of Fiscal Impact

Other cross-sector fiscal impacts arise from provision of low cost transit to low income groups by increasing accessibility of employment for these groups. Kain (1992) examined the spatial mismatch hypothesis in the context of housing policies. One of the interesting characteristics he discovered, was the unique composition of employment in the central Business District (CBD). Residents in and immediately surrounding the CBD are generally low- income and/or minority communities. However although these groups are within close proximity of the CBD, the majority of employees found therein are white, high-income earners brought in from the suburbs. "[T]he radial highway and transit systems in most large metropolitan areas make the CBD more accessible than many other parts of the central city to white suburban areas".[242] Kain (1992) asserts that if additional employment centers were made accessible by bus to the inner city, it would likely result in an increase in the percentage of jobs held by inner city residents in those areas. Consequently, this would decrease unemployment and social service expenditures.

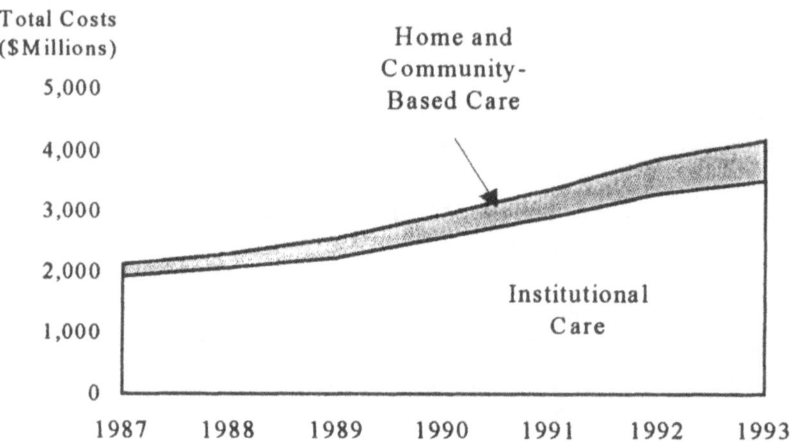

Figure 4.8 Medicaid Expenditures for Home and Community Based Institutional Care

Source: General Accounting Office. (August 1994). "Medicaid Long-Term Care: Successful State Efforts to Expand Home Services While Limiting Costs". *Report to the Chairman*, Subcommittee on Oversight and Investigations, Committee on Energy and Commerce, House of Representatives. p. 22.

Estimating Cross Sector Fiscal Impacts of Low Cost Mobility

This section presents a framework for operationalizing the estimation of cross sector fiscal impacts for the United States. Consistent with previous studies, cross sector fiscal impacts are defined to occur when mobility diminishes peoples' claim on other publicly subsidized assistance for "mobility surrogates" such ashome-delivered meals and health-care services. Transit, like any other public expenditure, must be considered in relation to governments' fiscal capacity. However, if in pursuit of spending reduction goals, a reduction in budgetary outlays for low cost mobility leads to a rise in entitlements and other budgetary expenditures on mobility surrogates, the fiscal framework is worsened rather than improved.

Table 4.10 Long Term Care Service Programs for the Aged and Persons with Physical Disabilities, 1993

Institutional Care	Services Provided	Average Monthly Users	State Fiscal Year 1993 Expenditures	Average Monthly Expenditure per User
Nursing Facilities	Personal care and services provided by licensed nursing personnel	17,428	$423,122,025	$2,023
Home and Community -based Care	Personal care, related household tasks, case management, supervision	22,040	$110,741,850	$419

Source: General Accounting Office. "Medicaid Long-Term Care: Successful State Efforts to Expand Home Services While Limiting Costs". *Report to the Chairman*, Subcommittee on Oversight and Investigations, Committee on Energy and Commerce, House of Representatives. (August 1994), p. 44.

The federal fiscal impact of transit and other mobility programs is both direct and indirect. The direct effect stems from the outlay of federal financial assistance for operating assistance and capital grants. These expenditures affect the federal deficit dollar-for-dollar (other things being equal) an additional dollar of transit expenditure translates into a dollar more on the federal deficit.

The indirect effects stem from the relationship between mobility-related program expenditures and expenditures on "mobility-vulnerable" non-transportation program expenditures in the federal budget. An elderly person using transit to gain access to a nutrition center, for example, creates a probability that he or she will make smaller claims on home-delivered meal services, such as the federally-assisted meals-on-wheels program, and other mobility-substitute services. Conversely, a cut in mobility precipitated by a cut in federal expenditure on transit creates a probability that mobility-vulnerable non-transportation expenditures, particularly entitlement expenditures, will rise to meet the demand for mobility-substitute services. The question addressed here is the risk that budgetary reductions in federal mobility related spending will generate increased rather than decreased fiscal outlays due to such cross-budgetary impacts.

Mobility-Vulnerable Federal Outlays

A review of the federal budget indicates that spending on programs that display some vulnerability to reduced mobility totaled $443 billion in 1994. Certain programs however are considerably more sensitive to the state of personal mobility than others. Focusing on just the three top-ranked programs from the viewpoint of their vulnerability to reduced mobility (Medicaid and Medicare, food support and unemployment insurance) yields total federal expenditures in 1994 of $293 billion.

The $293 billion in mobility-vulnerable program expenditures compares with total federal expenditures on mobility-related programs of $6.5 billion in the same year (this figure combines spending under the Federal Mass Transportation Act and spending under social service agency transportation provisions of the Older Americans Act).

Risk Analysis of Fiscal Impact From Reduced Mobility

Consider the scenario of a ten percent cut in federal expenditures on mass transit (including specialized transportation for people with disabilities) and social service agency-based transportation services. Since 60 percent of low cost mobility service users are without access to private automobiles, we assume that the ten percent reduction in federal spending diminishes mobility (trip making) six percent.

Estimation Methodology This section describes the step-by-step transportation and budgetary analysis underlying the estimates of the fiscal impacts of low cost mobility. Figure 4.9 presents a graphical illustration of this methodology, identifying all of the model inputs and the relationships between these inputs. As the figure indicates, the starting point assumes a

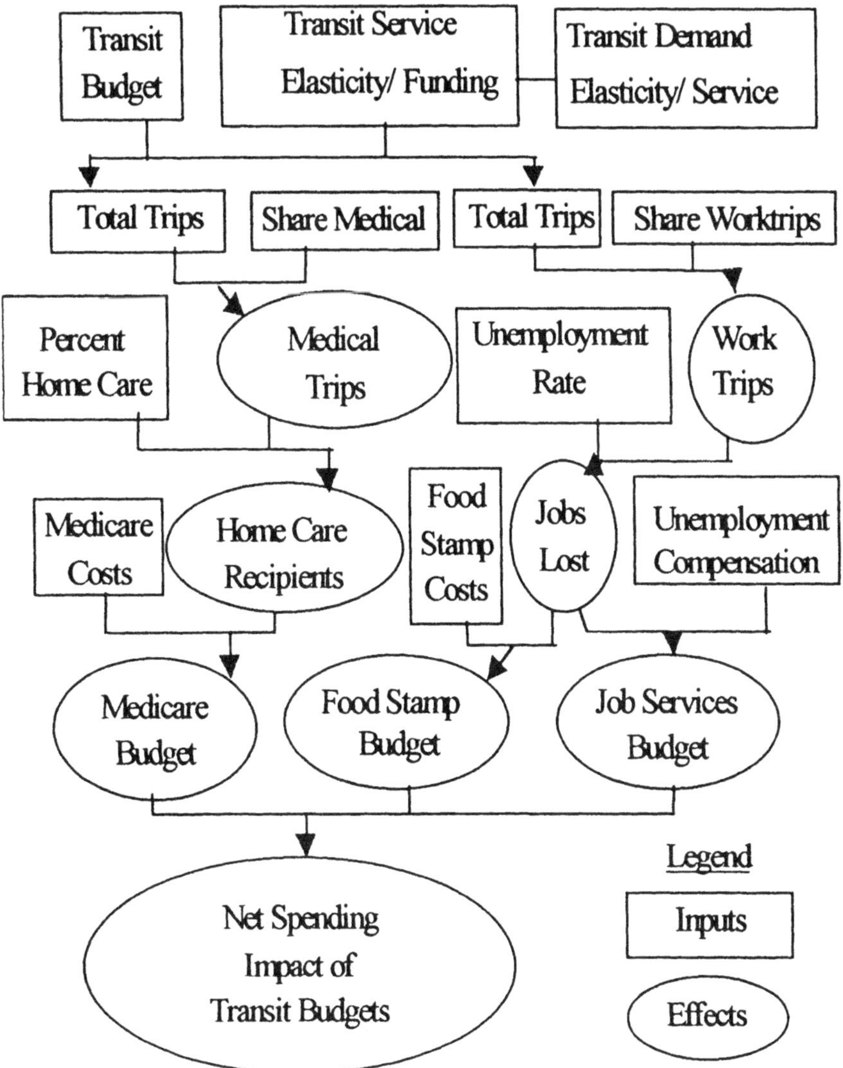

Figure 4.9 Method for Estimating Cross Sector Fiscal Impacts

change in transit funding. This change in transit funding is then translated into a change in service i.e., a change in transit demand. This second variable is expressed in elasticity terms i.e., service elasticity with respect to funding – the percentage change in trips associated with a one percent change in transit funding.

These first two steps result in the estimate of the change in the number of trips associated with the initial cut in transit expenditure and supply. This must be translated into trips by trip purpose in order to estimate fiscal impacts. The percent of lost medical trips which lead to home health care and lost work trips which lead to unemployment generate estimates of the number of added home care visits and number of lost jobs. The incremental Medicare-Medicaid program costs for each added home health care visit is multiplied by the number of added visits to estimate the monetary value of these trips. The same is repeated in relation to added food stamp costs and unemployment compensation whereby "benefits per lost job" is multiplied by the number of lost jobs to arrive at estimates of the monetary value of lost jobs. The risk analysis model is then simulated in order to obtain a probability range for the net budgetary impact of the ten percent cut in low cost mobility.

Programs Covered As a starting point for the estimation of the fiscal impacts from maintaining or improving mobility, we conducted a budgetary analysis and cross referenced this analysis with the data on public transportation trips by trip purpose.

Table 4.11 presents the results. Mobility vulnerable programs fall into four categories: Health, Employment, Education, and income support. These programs accounted for $443 billion in federal spending, or about 30 percent of the total budget. This analysis focuses on four particularly vulnerable programs: Medicaid, Medicare, Food Stamps and Unemployment Compensation. They accounted for $293 billion in 1994.

Data And Technical Assumptions To operationalize this framework, we conducted research to populate the model framework. Table 4.12 presents model inputs and probability ranges for each of the model variables described in Table 4.11.

Cross-Sector Fiscal Impact Results

Table 4.13 presents the results of the risk analysis, by program, in terms of the expected value (i.e., 50 percent probability of at least this value). Medicare/Medicaid program costs are expected to increase by $162 million as a result of a $650 million dollar cut in federal mobility related programs. Food Stamp and unemployment insurance programs are expected to account for an additional $233.6 million in additional spending as a result of the $650 million dollar budgetary reduction. In total, the cross sector fiscal impact as a result of the reduced expenditure is expected to be $395.7 million. This indicates that for a one dollar reduction in transit budget there will be a net savings of only $0.39.

Table 4.11 Mobility Vulnerable Programs Ranked From Most to Least Vulnerable, 1994

Federal Mobility Vulnerable Programs	Funding (Billions)
1. Medicaid	$82
2. Medicare	$160
3. Food Stamps	$25
4. Unemployment Compensation	$26
5. Social Services	$6
6. Supplemental Security Income	$24
7. Family Support	$17
8. Earned Income Credit	$11
9. Federal Civilian Retirement & Disability	$40
10. Military Retirement & Disability	$27
11. Veteran's Benefits	$18
12. Child Nutrition	$7
Total Spending on Mobility Sensitive Programs	$443

Source: Federal Budget of the United States, 1994.

Table 4.14 summarizes the risk analysis results. The analysis shows that the federal government faces a 90 percent risk of precipitating a $106.5 million rise in federal spending on safety-net services as a results of a ten percent cut in federal funding for low cost mobility programs.

Alternatively, the government could suppress some or all of this increase in entitlements and other program outlays. The latter outcome is likely if only for reasons of limited information about cross-sectoral effects rather than policy priorities. Under this, the most likely outcome, *both* public transit users and users of safety-net programs stand to lose from transit cuts. Conversely, transit can be seen to create significant club and spillover effects in a nation's economy.

Table 4.12 Risk Analysis Model Inputs With Probability Ranges

Model Input	Constant or Median	Lowest Decile	Highest Decile
Transit Trips (bn)	7.2		
Transit Funding (bn)[a]	$6.5		
Percent Change in Transit Funding [a]	10		
Elasticity of Service w.r.t Funding[b]	1.8	0.8	2.5
Elasticity of Demand w.r.t Service[b]	1.0	0.4	1.5
Percent of Trips for Medical Purpose[c]	5		
Percent of Trips for Work Purpose[c]	54		
Percent of Medical Trips by Medicare/Aid Recipients[c]	80	75	85
Percent of Lost Medical Trips That Go to Home Care[d]	10	5	15
Incremental Cost of Home Care ($)[e]	$50	$45	$55
Percent of Lost Work Trips Leading to Unemployment[f]	3	2	3.2
Avg. Wage of Newly Unemployed	$6/hr	$5/hr	$6.50/hr
Food Stamp Program Annual Cost/Recipient	$1,428	$1,285	$1,570

[a] Hickling Lewis Brod, Economics estimates based on the Federal Budget of the United States of $6.5 billion Includes $4.5 billion in FTA appropriations, $0.8 billion in flexible spending shifted to transit purposes, $1 billion in transit related Medicaid spending and $0.2 billion in Department of Health and Human Services budget dedicated to transit spending.

[b] Hickling Lewis Brod Economics estimates.

[c] Section 15 data and the NPTS 1994.

[d] Hickling Lewis Brod, Economics estimates based on Health Care Financing Administration estimates of medical trips and the National Home and Hospice Care Survey. 2 billion total home health care visits are covered by Medicare/Medicaid.

[e]Based on the total cost of a home health visit of $74 less the amount of spending that would occur otherwise (assumed to be approximately $25). Statistics provided by the Office of the Actuary, Health Care Financing Administration and the National Home and Hospice Care Survey.

[f]Hickling Lewis Brod Economics estimate.

Table 4.13 Fiscal Impacts, by Major Program of a 10 Percent Transit Funding Reduction, Based on 1994 Program Budgets

Costs by Affected Program	1994 Budget Allocation ($Millions)	Estimate of Fiscal Impacts ($Millions)
Medicare/Medicaid Program	$242,000	$162.1
Food Stamp Program	$25,000	$64.5
Unemployment Insurance	$26,000	$169.1
Fiscal Impacts	$293,000	$ 395.70

Source: Hickling Lewis Brod Economics estimates and *The Federal Budget of the United States*, 1994.

Table 4.14 Transit Impacts on Collateral Social Budgets

Increase in NonTransportation Program Expenditures ($ millions)	Probability of Exceeding Value to the Left
59.5	95%
106.5	90%
186.5	80%
321.1	60%
395.7	50%
505.7	30%
696.9	10%
791.8	5%
Mean = 395.7	Standard Dev. = 223.8

Source: Hickling Lewis Brod Economics Analysis.

Summary

This chapter evaluates four alternative frameworks for considering the benefits of low cost mobility. The first is based on a longstanding paradigm in public finance referred to as Merit Goods. While on ideological grounds, the standard merit good argument for supporting publicly provided or subsidized programs may be valid, it is difficult if not impossible to validate empirically. The second framework of income distribution is also a mainstay of public finance analysis. But, it is difficult to determine the efficacy of income distribution programs and the analytics of this framework provide no insight as to the benefits of low-cost mobility as compared to the costs.

The two remaining approaches are based on the economic and fiscal "value" of low cost mobility. Both of these frameworks can be readily implemented using widely accepted techniques of economic and budget analysis. These approaches are also supported by accessible and credible data sources.

The first of these two approaches is based on the basic principles of cost-benefit analysis where the economic value people obtain from low cost mobility is the value they derive from their journey purposes (work, shopping, going to school etc.) minus the fares they pay. In 1993, the economic value of transit as a source of low cost mobility was $33.7 billion. The net economic return to the nation from its investment in low cost mobility was an estimated to be $17.5 billion in 1993 ($33.7 billion in the value obtained by individuals minus the $16.24 billion in nation-wide public expenditures to operate, maintain and invest in transit systems).

The final approach generates an estimate of the cross-sector fiscal impacts associated with a $650 million (10 percent) reduction in the federal budget for low-cost mobility programs. This analysis reveals an expected impact of this reduction on the four most mobility vulnerable programs: Medicare, Medicaid, Food Stamps and Unemployment Insurance, of approximately $396 million. Or, for every dollar saved with the mobility program cut, there is a net savings of $0.39 because of the increased costs in mobility-vulnerable programs.

Annex 4.1 Sensitivity Analysis for Estimates of the Economic Value of Affordable Mobility

Sensitivity analysis is conducted to account for the uncertainty surrounding the key parameters at the low income trip point, namely elasticity with respect to fare, and the corresponding willingness to pay. This is achieved by identifying feasible pairs of elasticity and fare which fulfill the following: The demand schedule (Figure 4.10) does not violate the Arrow–Debreau general form of demand curve, goes through the "known" point were the elasticity is -0.3, the trip level is 7.2 billion, and the fare is $1.50.

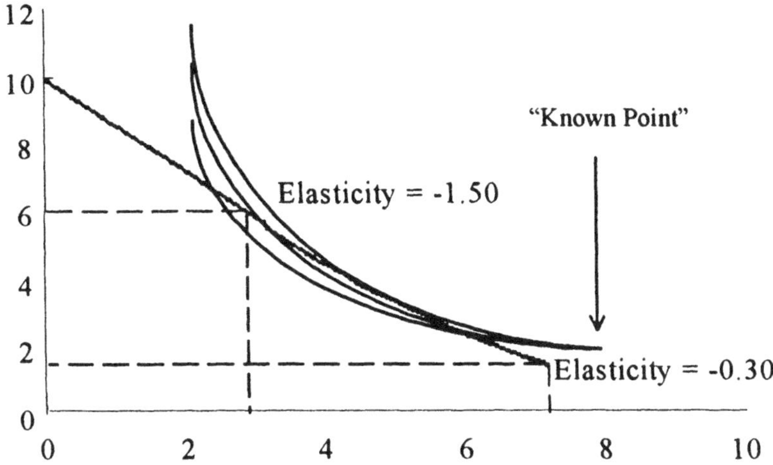

Figure 4.10 Sensitivity of Demand Schedule

Source: Hickling Lewis Brod Economics, Inc., 1996.

The consumer surplus and the maximum willingness to pay of low income individuals are re-estimated, varying the assumptions for elasticity and willingness to pay at the low income threshold.

The ranges of elasticity and willingness to pay used for the sensitivity analysis are based on indications of previous studies. The ranges used are those that most likely reflect the willingness to pay for transit trips by low income individuals.

Results The following series of tables identify the feasible pairs of elasticity and willingness to pay by low income individuals calculating the consumer surplus and maximum willingness to pay for all transit users (using $20K as the low-income anchor point), and separately for each income category of interest.

As indicated in Table 4.15, the lower bound consumer surplus is approximately $21 billion. This point is the minimum consumer surplus estimated without the assumption that low income users are more responsive to change in price. The maximum willingness to pay by low income individuals at this point is $7.50.

The estimate of consumer surplus under the original assumptions, lies approximately in the middle of the range at $33.5 billion. The maximum willingness to pay by low income individuals at this point is $12.6. The high bound of the sensitivity analysis indicates a consumer surplus of approximately $64 billion. The maximum willingness to pay by low income individuals at this point is $28.00. NA indicates a non-feasible combination of price and elasticity, that violates the Arrow-Debreau general form of demand curve as indicated above.

The following tables show the sensitivity analysis for the consumer surplus and for the maximum willingness to pay by individuals of different income categories that are of interest: low income households (Table 4.16), households below Poverty Line (Table 4.17), household below the Earned Income Tax Credit Ceiling (Table 4.18), and households below the U.S. median annual income (Table 4.19).

Table 4.15 Consumer Surplus and Maximum Willingness-To-Pay for All Income Groups

"Feasible" Combinations of Price and Elasticity*

($ Billions)

		Willingness to Pay ($)							
		5	6	7	8	9	10	11	12
Price Elasticity for Low Income Patrons	-0.75	26.21	32.38	38.31	44.00	49.45	54.67	59.65	64.38
	-1	23.96	29.70	35.21	40.48	45.51	50.30	54.86	59.18
	-1.25	22.67	28.17	33.45	38.49	43.29	47.86	52.18	56.27
	-1.5	21.85	27.22	32.37	37.27	41.94	46.37	50.56	54.52
	-1.75	21.31	26.60	31.66	36.49	41.08	45.43	NA	NA
	-2	20.95	NA	NA	NA	NA	NA	NA	NA
	-2.25	NA	NA	NA	NA	NA	NA	NA	NA

NA indicates non-feasible combination of price and elasticity.

* Evidence indicates that the actual value lies in the outlined area of the table.

Table 4.16 Consumer Surplus and Maximum Willingness-To-Pay for Low Income Individuals (Below $20 Thousand Annual Household Income)

"Feasible" Combinations of Price and Elasticity

($ Billions)

		Willingness to Pay ($)							
		5	6	7	8	9	10	11	12
Price Elasticity for Low Income Patrons	-0.75	19.68	24.48	29.28	34.08	38.88	43.68	48.48	53.28
	-1	17.28	21.60	25.92	30.24	34.56	38.88	43.20	47.52
	-1.25	15.84	19.87	23.90	27.94	31.97	36.00	40.03	44.06
	-1.5	14.88	18.72	22.56	26.40	30.24	34.08	37.92	41.76
	-1.75	14.19	17.90	21.60	25.30	29.01	32.71	NA	NA
	-2	13.68	NA	NA	NA	NA	NA	NA	NA
	-2.25	NA	NA	NA	NA	NA	NA	NA	NA

NA indicates non-feasible combination of price and elasticity.

Table 4.17 Consumer Surplus and Maximum Willingness-To-Pay for Below Poverty (Below $11.5 Thousand Annual Household Income)

"Feasible" Combinations of Price and Elasticity*

($ Billions)

		Willingness to Pay ($)							
		5	6	7	8	9	10	11	12
Price Elasticity for Low Income Patrons	-0.75	NA	12.75	15.25	17.75	20.25	22.75	25.25	27.75
	-1	NA	11.25	13.50	15.75	18.00	20.25	22.50	24.75
	-1.25	NA	10.35	12.45	14.55	16.65	18.75	20.85	22.95
	-1.5	NA	9.75	11.75	13.75	15.75	17.75	19.75	21.75
	-1.75	NA	9.32	11.25	13.18	15.11	17.04	18.96	20.89
	-2	NA	9.00	10.88	12.75	14.63	16.50	18.38	20.25
	-2.25	NA	8.75	10.58	12.42	14.25	16.08	17.92	19.75

NA indicates non-feasible combination of price and elasticity.

* 11.5K is the poverty line income level for a 3.2 person household.

Table 4.18 Consumer Surplus and Maximum Willingness-To-Pay for Individuals Below the Earned Income Tax Credit Ceiling (Below $24.5 Thousand Annual Household Income)

"Feasible" Combinations of Price and Elasticity*

($ Billions)

		Willingness to Pay ($)							
		5	6	7	8	9	10	11	12
Price Elasticity for Low Income Patrons	-0.75	23.92	29.75	35.58	41.42	47.25	53.08	58.92	64.75
	-1	21.00	26.25	31.50	36.75	42.00	47.25	52.50	57.75
	-1.25	19.25	24.15	29.05	33.95	38.85	NA	NA	NA
	-1.5	18.08	NA	NA	NA	NA	NA	NA	NA
	-1.75	NA	NA	NA	NA	NA	NA	NA	NA
	-2	NA	NA	NA	NA	NA	NA	NA	NA
	-2.25	NA	NA	NA	NA	NA	NA	NA	NA

NA indicates non-feasible combination of price and elasticity.

* 24.5K is the earned income credit level for 3 person household.

Table 4.19 Consumer Surplus and Maximum Willingness-To-Pay for Individuals Income Below the National Mean (Below $34.5 Thousand Annual Household Income)

"Feasible" Combinations of Price and Elasticity*

($ Billions)

		Willingness to Pay ($)							
		2	3	4	5	6	7	8	9
Price Elasticity for Low Income Patrons	0	NA	NA	NA	NA	NA	NA	NA	NA
	-0.25	NA	NA	50.4	64.8	79.2	93.6	108.0	122.4
	-0.5	NA	NA	31.2	40.8	50.4	60.0	69.6	79.2
	-0.75	NA	NA	24.80	NA	NA	NA	NA	NA
	-1	NA	NA	NA	NA	NA	NA	NA	NA
	-1.25	NA	NA	NA	NA	NA	NA	NA	NA
	-1.5	NA	NA	NA	NA	NA	NA	NA	NA

NA indicates non-feasible combination of price and elasticity.

* $34.5k is the Median Urban Household Income.

Notes

188 Musgrave, R.A., *The Theory of Public Finance,* New York: McGraw-Hill, 1959); Musgrave developed the theory of merit goods based on the original work of Pigou.

189 Hickling Lewis Brod Economics, "Basic Mobility Needs and Related Transit Activities" FTA Working Paper 1.1 (July 27, 1995). Ref: 5273-300, p. 52.

190 Corresponding to Aristotle's ultimate good, the *telos*. *The Politics.*

191 Mayer Hillman, Irwin Henderson, Anne Whalley, "Personal Mobility and Transport Policy", PEP, Vol. 34, Broadsheet 542 (London, June 1973), p. 134.

192 Ibid., p. 52.

193 Kenneth J. Button, *Transport Economics--2nd Edition*, (Vermont: Edward Elgar, 1993), p. 50.

194 Ibid., p. 132

195 Alex Meyerhoff, Martine Micozzi and Peter Rowen, "Running on Empty: Travel Patterns of Extremely Poor People in Los Angeles", *Transportation Research Record* 1395, (Washington, D.C.: Transportation Research Board, 1993), p.160.

196 Tax/expenditure literature.

197 Button, op. cit., p. 51.

198 Bellah, Robert N., Richard Madsen, William M. Sullivan, Ann Swidler, and Steven M. Tipton, *Habits of the Heart: Individualism and Commitment in American Life*, (New York: Harper and Row, 1985).

199 Mayer, et al, op. cit., p. 9.

200 Ibid., p. 153.

201 Ibid., p. 158.

202 Kain studied the effect of housing market discrimination on the employment and earnings of African-American workers. He hypothesized that residential segregation affects the employment aggregate level of demand for this group of workers.

203 John F. Kain, Housing *Policy Debate--The Spatial Mismatch Hypothesis: Three Decades Later*, (Washington, D.C.: Fannie Mae, 1992), p. 371-459.

204 Mark A. Hughes, *Fighting Poverty in Cities: Transportation Programs as Bridges to Opportunity.* (Washington, DC: National League of Cities, 1989), pp. 203-204.

205 Based on statistics from the Nationwide Personal Transportation Survey, 1990.

206 David Lewis, *The Economics of Serving the Travel Needs of Handicapped Persons in the United States*, (London: London School of Economics, 1985).

207 J. Pucher and F. Williams, "Socioeconomic Characteristics of Urban Travelers: Evidence from the 1990-91 NPTS", *Transportation Quarterly*, Vol. 46, No. 4, (October, 1992).

208 Ibid.

209 One might question the proposition that a journey's value to the traveler always exceeds its price. Consider trips made by very low wage earners for whom the cost-of-living (including travel) exceeds total earnings plus any supplementary government financial assistance, for example. The problem however is one of semantics. When those who sustain an activity pattern like that in the example are said to "value" intangible attributes of work over the goods and services they need to forgo in order to remedy their cost-of-living deficit, it would be adequate to replace the word "value" with the more neutral term "choose". Indeed, in many such cases people are compelled to choose from competing "bads" (work at a loss or sustain the indignity of welfare, for example) rather than from competing preferences and the idea that they "value" the choice they make seems callous in this context. The fundamental proposition however remains valid; when trips are made, it must logically be assumed that the benefits derived by the traveler, as he or she perceives them, exceed the costs.

210 Meyerhoff, et al, op. cit. p.160

211 Ibid., pp. 154-155.

212 Hank Dittmar and Don Chen, "Equity in Transportation Investments", *Environmental Justice and Social Equity Conference*

Proceedings, (Washington, D.C.: Transportation: Federal Transit Administration, 1995), p. 40.

213 Mayer, et al, op. cit., p. 123.

214 John Pucher, "Discrimination in Mass Transit", *Journal of the American Planning Association*, (Summer 1982), pp. 315-326. See also Don Chen. "Social Equity, Transportation, Environment, Land Use, and Economic Development: The Livable Community", *Transportation: Environmental Justice and Social Equity Conference Proceedings*. (Washington, D.C.: Federal Transit Administration, July 1995), pp. 42-47.

215 Ibid., pp. 315-326.

216 Susana Almanza and Raul Alvarez. (July 1995). "The Impacts of Siting Transportation Facilities in Low-Income Communities and Communities of Color". *Transportation: Environmental Justice and Social Equity Conference Proceedings*. (Washington, D.C.: Federal Transit Administration, July 1995), p. 34.

217 Mark Roseland, *Toward Sustainable Communities--a Resource Book for Municipal and Local Governments*, (Ottawa, Canada: National Round Table series on Sustainable Development, 1992), p. 250.

218 Dittmar and Chen (July 1995), op. cit., p. 38.

219 See Section-April 1996 The Economic and Fiscal Value of Low Cost Mobility-p. 46.

220 The Consumer Expenditure Survey for 1994 indicates that the poorest households spend on travel by automobile 25 percent as much as high income earners.

221 Federal Highway Administration, Nationwide Personal Transportation Survey, 1990.

222 David Lewis, op. cit.

223 Ibid. and United States Department of Transportation, Nationwide Personal Transportation Survey, 1990.

224 Ibid.

225 "ADA" denotes Americans with Disabilities Act. ADA regulations do not require the public provision of door-through-door or gurney-accessible transportation.

226 David Lewis, op. cit.

227 Unweighted average fare across all modes (including commuter) is approximately $1.00 (1995). When this fare is weighted according to the number of trips taken in each jurisdiction, the average fare increases to approximately $1.50 . APTA, *1996 Transit Fact Book*, (Washington, D.C.: American Public Transit Association, May 1996).

228 For some people of course willingness to pay will understate the true value they confer on a trip purpose simply because their ability to pay is severely constrained by economic circumstances. This issue is addressed later.

229 The average fare is an Hickling Lewis Brod Economics estimate based on data provided by APTA. It is an estimate of a representative fare actually paid by users, rather than fare revenue per trip as is often quoted. As such, Hickling Lewis Brod Economics adjusted for the number of trips which were not covered by fares and estimated the fare value by taking the distribution of trips by region, collecting data on actual fares per trip and estimating a weighted average fare.

230 See David Lewis, "Public Transport Fares and the Public Interest", *Town Planning Review*, Vol. 46, No. 3, (July 1975).

231 Melanie Carr, Tim Lund, Philip Oxley and Jennifer Alexander, *Cross-sector Benefits of Accessible Public Transport,* (Crowthorne, Berkshire, U.K.: Environment Resource Centre, 1993), p. 1.

232 Ibid., p. 4. Nominal figures have been adjusted using the 1993 United States / UK exchange rate of 1.5016. Article stated benefits of 30,000 and 40,000 pounds respectively.

233 Ibid.

234 Carr et al., op. cit., p.9; *Managing Social Services for the Elderly More Effectively*. (London : HMSO Audit Commission, 1985).

235 Carr et al., op. cit., p. 12.

236 Ibid., p. 15.

237 Ibid., p. 15-23, for complete methodology and results. Also, benefits reported in our discussion are quoted in United States $; 1993 exchange rate of 1.5016 used.

238 Ibid., p. 25.

239 For complete explanation of methodology, see ibid., pp. 43-49.

240 Ibid., p. 27.

241 "Medicaid Long-Term Care: Successful State Efforts to Expand Home Services While Limiting Costs". *Report to the Chairman*, Subcommittee on Oversight and Investigations, Committee on energy and Commerce, House of Representatives, (Washington, D.C.: General Accounting Office, August, 1994), p. 2.

242 Kain (1992), op. cit. p. 397.

5 Transit Value to Neighborhoods

Introduction

For many Americans, living near high quality rail transit stations provides an array of benefits. These benefits arise from lower transportation expenses, the conveniences of ready access to myriad services, and other non-user benefits. The purpose of this chapter is to explore the benefits of transit oriented neighborhoods using economic measurements.

This chapter employs a hedonic price function to estimate property values and the impact of proximity to rail transit stations. We used Geographical Information System (GIS) databases to calculate actual walking distances to transit. This methodology provides a much more accurate measure of the "proximity" variable than the usual measure of straight line distance.

In this chapter, the value of transit is estimated in three areas: San Franciso's Bay Area Rapid Transit (BART) station in Pleasant Hill, New York City subway stations in Queen, and light rail stations along the East Burnside corridor of Portland, Oregon.

The results of the analysis for the BART station in Pleasant Hill and the Queens Borough of New York City, show strong property value impacts relating to station proximity. In fact, the values found in the hedonic model results likely exceed those attributable to time savings alone. The existence of non-user benefits is the most likely explanation.

Historical Urban Paradigms

This review presents and critiques the history of urban development paradigms. It assesses "new urbanism" and other approaches for livable communities and traces the link with Transit-Oriented Development (TOD).[243] Further, it considers state of the art analytical procedures and techniques used to study the attributes of livable neighborhoods and reports the results of past research valuing the benefits of livability and access. Studies of transportation's role in urban neighborhood development diverge in many respects, but tend to agree on three conclusions that are important for this chapter:

Current auto-dominated development patterns are creating negative impacts manifested as congestion, pollution, and urban sprawl;

Transit-oriented communities resolve many of these negative impacts; and

Property value models are the most reliable means of estimating development benefits from transportation investments.

Historical Development of Neighborhood Ideals

Historical and social developments affect the physical and social character of neighborhoods (see Figure 5.1). The attributes of neighborhoods reflect people's belief systems and values. This chapter will show how past metropolitan development ideals, often influenced by government intervention, failed to produce "livable" neighborhoods. It concludes with an introduction to the "new urbanism" and its relation to transit-oriented development.

The concept of a "livable" neighborhood eludes simple definition by a fixed group of physical, social and environmental attributes. The term "livable community" has been used as shorthand for alternative development to auto-dominated suburbs. It is, however, the descriptor that has been chosen in the urban planning and architectural literature. The subsequent research effort is intended to estimate the value of "livable" communities, expressing their "livability" in quantitative terms.

Every community depends upon the interaction of local physical resources and human talent to develop the collective vision of the community, and every community contains and lacks attributes that some describe as "livable community" attributes. In urban areas, livable neighborhoods might take the form of compact, high density, mixed-use, contiguous urban development as suggested in much of the literature,[244] while rural livable communities might have far different characteristics. This review focuses on the urban livable community and the range of common attributes associated with it.

Urban planners and architects, local government officials, politicians and academics, have placed the promotion and development of "livable" neighborhoods at the center of many current public policy debates.

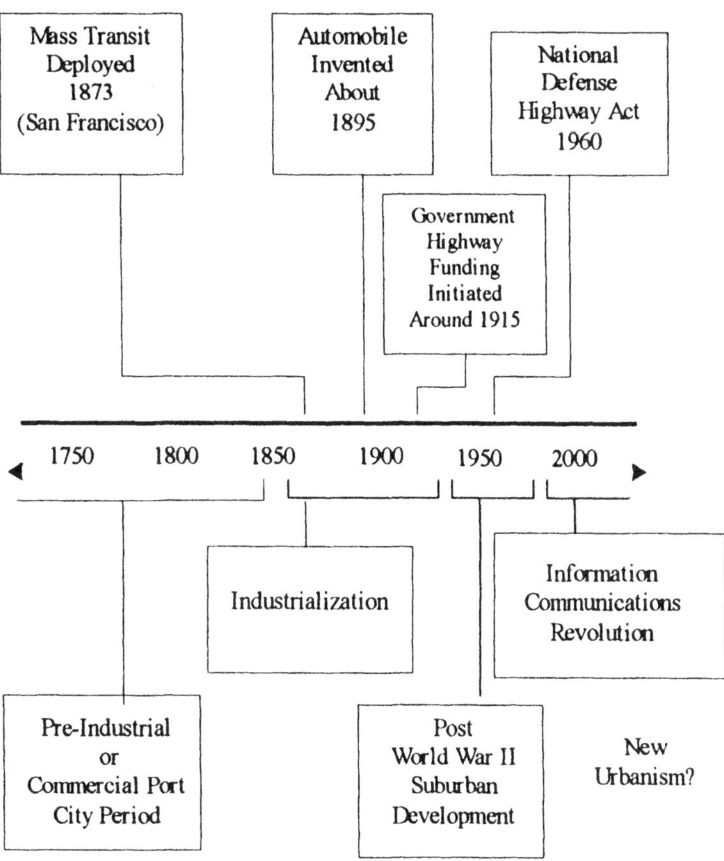

Figure 5.1 Time Line of Major Historical Development Affecting Transportation and Urban Development

Whether based on economic, social or environmental grounds, livable neighborhoods confer a set or bundle of common attributes to residents. This economic or social value of these attributes may be intuitively evidenced by virtue of property price premiums and the willingness of citizens to protect and promote their collective ideals or vision. The linkage between the attributes of a livable neighborhood and the intrinsic values that it confers is the foundation of the quantitative research of the current project.

This review draws on literature from several disciplines to develop a comprehensive classification of key neighborhood attributes that residents and specialists alike ascribe to contemporary, urban livable communities. Given the vast diversity of communities and the range of urban forms, different combinations of these attributes may define a livable community. As this review shows, North American cities have undergone massive changes throughout their histories in response to changes in socio-economic trends and evolving visions of neighborhood ideals. These changes have brought great improvements in the quality of most Americans' lives but also created some unintended consequences. Whether new visions of ideal neighborhood developments alleviate those consequences while maintaining historical gains is the goal of this review.

Livability and Urban Form

Urban communities include metropolitan areas and are not limited to central cities. Urban problems and benefits affect suburban areas as well as center cities. Therefore, "livable" neighborhood attributes and the means of achieving benefits should apply to all communities in a metropolitan area. In fact, the emergence of the suburban ideal appears to be a reaction against the poor living conditions in the industrial city. The newer ideals of Transit-Oriented Development can be seen as a reaction against the sprawl and dislocation of the current suburban style city.

There are many factors that may make urban living difficult, including: traffic congestion, pollution, inconvenience, lack of affordable housing, crime, bad schools, high taxes, and inadequate public services. Some authors have posited that many of these urban problems have their root in the manner in which we build cities. In essence, the infrastructure that is built, the zoning regulations that control private development, and the tax policies that influence growth patterns contribute, in some way, to the problems plaguing urban areas.

Some characteristics of urban areas such as long commutes, congestion, pollution, and lack of convenience are most directly influenced by transportation infrastructure design and investment. The idea that the type of transportation infrastructure we build, from streets to transit facilities, can bring about benefits in terms of "livability" has come to embody the ideas of Transit-Oriented Development.[245] This review presents Transit-

Oriented Development as a branch of the "new urbanism" and presents its guiding principles.

Historical Context of the Livable Neighborhood

The livable neighborhood concept varies significantly among academic disciplines and across time. The architect Frank Lloyd Wright, working in the first half of the 20th century, stressed the individual's close integration with nature through detached, single unit housing. This ideal is contrary to the ideal, supported by environmentalists such as Marcia Lowe, of an environmentally friendly "livable city" with multi-unit housing, mixed-land use, planned green space and bicycle and pedestrian paths.[246] The difference between Wright's and Lowe's concept of the livable community arose from their ideological differences and the period during which they worked.

The ideal of a "livable" community is affected by prevailing political, social, economic and physical factors. This review surveys the history of urban development in North America to understand how our nation came to be dominated by low density suburban development. America's development patterns have a number of precedents and influences as well as some distinct ties to public policies and technology. The following sections present this history.

The Pre-Industrial Cities

New York was founded in 1623–only fifteen years after Quebec—but by 1664, when the Dutch handed over what was then called New Amsterdam to the British, the walled town contained 10,000 persons. The influence of old Amsterdam was evident in the canal that led to the center of town, the brick construction, and the gable-fronted houses; behind the houses were large garden plots and orchards.[247]

The first generation of American cities reflected the traditional English preference for casual planning and improvisation. The layouts of pre-industrial, colonial towns followed three general patterns:

Angled and winding streets growing informally over a period of time. Boston is an example of this pattern.

> Gridded plans interspersed with occasional open squares. Examples of this style town are Cambridge, Massachusetts and Hartford, Connecticut.
>
> Linear towns organized along a main street. An early example is Providence, Rhode Island.

Many towns combined two or more of these patterns. As towns expanded and grew into larger cities, the initial street patterns were typically expanded (Rybczynski, p. 67).[248]

A defining element of the pre-industrial city was the extent to which land uses were mixed together. Homes, places of business and warehouses were next to each other, and sometimes, were the same building. Business and social interaction occurred in face to face interactions in public areas or taverns, and people from all backgrounds mixed in many different settings. Of course the pre-industrial city was limited in geographical area by the necessity for a city's functions to be within walking distance.

The mixed-use nature of pre-industrial cities was not always pleasant. Streets were not always paved, sewers were open if they existed at all, and transportation was by foot or horse. As a result of these limitations, cities were not usually large during this period. By 1831, Cincinnati had grown to about 30,000; Washington contained fewer than 20,000; New York had reached a *massive* 200,000 while Baltimore was a burgeoning city of 80,000.[249] The size of the pre-industrial city was limited by the inability of the urban technologies and transportation systems to support much larger cities.

In 1831, pre-industrial New York was an unpleasant place with little in common with the quote provided above. Large sections of the city were destroyed by fire during the British occupation. The city recovered rapidly, however, to become the largest and busiest city in the nation. As a bustling pre-industrial city with no underground sewers, the waste from 200,000 people and the animals they used for transportation was either collected by night-soil scavengers and carted to the countryside for fertilizer or was deposited into cesspools contributing to the contamination of the city water supply. As Rybczynski writes:

> As [Alexis De] Tocqueville noted, most streets were unpaved and crowded with thousands of wagons and carriages. The mud in the street was mixed with horse manure, and domestic waste was scattered everywhere, for there was no trash collection. Garbage simply accumulated outside and was trampled into the street, which explains why the oldest Manhattan streets are anywhere from six to fifteen feet

higher than their original levels. Scavenging pigs wandered the streets and sidewalks. There was no mass transportation, few building regulations and such poor fire protection that a great fire destroyed more than fifty acres in lower Manhattan only three years after Tocqueville's departure. Life in the big city was dangerous, uncomfortable, and unhealthy.[250]

During the pre-industrial period, America was first and foremost an agrarian society with a relatively small proportion of the population living in cities for economic, social, and technological reasons. The industrial revolution radically altered the character of the pre-industrial city, mainly by drawing the population from the countryside to urban areas and into what H.G. Wells dubbed "whirlpool cities".[251]

Industrial Cities

> "Evil conditions were to be found in every section of the city [Pittsburgh, PA]. Eyrie rookeries perched on the hillsides were swarming with men, women and children...entire families living in one room and accommodating "boarders" in a corner thereof. Courts and alleys fouled by bad drainage and piles of rubbish were playgrounds for rickety, pale-faced children. An enveloping smoke and dust through which light and air must filter intensifies the evil of overcrowding..."[252]

The industrial period saw rapid urbanization of American society, drawn from rural areas as well as from abroad in large waves of immigration. The cities that were moderate sized pre-industrial cities in the early nineteenth century had become much closer to the metropolises we know today. In 1892, Chicago had more than 1.5 million residents, New York had exploded to over 3.4 million, while Philadelphia stood at over 1.3 million.[253] The industrial cities signaled the shift of the American population from rural to urban. In 1831, only 10 percent of the population lived in cities while by 1892, the figure was nearly 40 percent (see Figure 5.2).

Early Industrial Cities

The industrial city encompassed two trends affecting the livability of urban areas. First, the massive influx of population strained the fabric of society producing conditions of dense and dirty tenements. Second, urban technologies developed in response to this growth including: modern roads, sewers, public transportation and other infrastructure allowed the development of a much more "livable" city. The early industrial cities of the late nineteenth century suffered from generally poor and dirty living conditions. The cities were subject to overwhelming inflows of people while the urban technologies to handle the influx could not keep pace or were not developed.

The resulting pollution and overcrowding created the impetus for a new vision of livability. This enormous expansion of economic opportunity, as it created dismal conditions in many urban areas, also produced a massive expansion of wealth. This wealth provided the means for improving the livability of cities and, combined with improvements in transportation systems, allowed the exploration of new visions for urban living. In essence, the suburb was born.

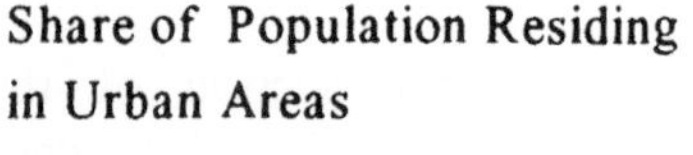

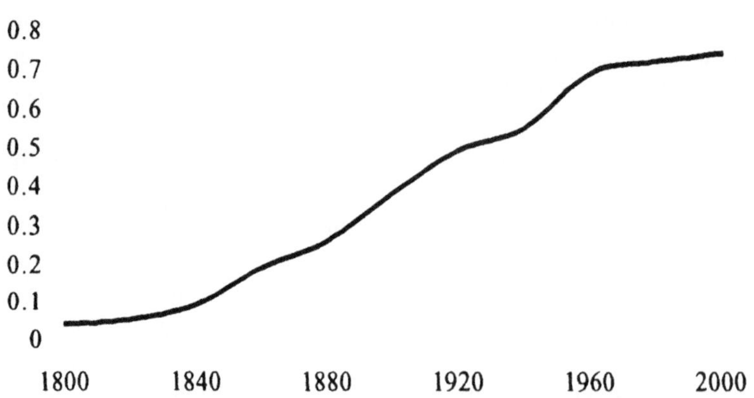

Figure 5.2 Urbanization in the United States, 1800 - 2000

Source: Sullivan, Arthur, *Urban Economics*, (Boston: Irvin, 1993), p. 88.

Searching for "Livable" Cities

The view in the early part of this century that the ideal community should include lower density development, separated land uses, and free standing buildings surrounded by open space was largely a reaction to the rapid urbanization and degrading conditions in 19th century newly industrial cities.[254] This desire for separation and control of land use and the concurrent advances in transportation technology such as the streetcar followed by the automobile, led directly to the desire for new residential developments outside of central cities. And, subsequently, a demand for automobile friendly road networks to connect the new developments to the spatially separated commercial or industrial districts.

The transportation-related technological innovations had the effect of increasing the land area within commuting distance of employment centers. The transportation innovations that made this expansion possible began with the streetcar and the subway in the last part of the 19th century. Many United States cities are, in fact, distinguished by old "streetcar" suburbs where newly mobile citizens began to experience the "American dream" of life in the suburbs. These "streetcar" suburbs however had a very different character than the automobile dominated suburban development that has occurred after World War II. The streetcar suburb needed to maintain "walkability" since cars were not widely owned until well into the 20th century. The need to remain within moderate walking (or horse) distance of stores, services, and the streetcar, helped to encourage higher density development than current patterns and maintained conditions amenable to walking and public transit. This late 19th century and early 20th century suburban development pattern is one inspiration for the New Urbanism and Transit-Oriented Development; a modern vision of the streetcar suburb interpreted in urban and suburban settings.

Transportation and the Government

Neighborhood form is strongly influenced by the availability of transportation infrastructure. The preponderance of the streetcar and the relative lack of automobiles ensured that the new suburbs were pedestrian oriented and close to transit.

The majority of transit systems in the early part of this century were privately owned and operated. They were profitable, received no public support, and were heavily used. Transit was burdened, however, with extensive fare regulation by local governments and were not allowed to charge higher fares despite rapidly rising costs after World War I. These pressures eventually backrupted nearly one-third of the nation's streetcar companies. Low fares and high costs starved the transit industry of capital to fund infrastructure improvement and expansion.[255]

The rise of the automobile and the decline of transit began during this period and the driving force was massive intervention by the government on the automobile's behalf on top of excessive fare regulation. Beginning in the 1920s, government agencies were pouring $1.4 billion per year into highways. By 1940, government was spending $2.7 billion on highways rising to $4.6 billion in 1950 and exploding to over $11 billion in 1960 with the National Defense Highway Act. This level of support for automobiles virtually eliminated the financial viability of private transit companies.

Post War Expansion and Auto-Oriented Suburbs

> "A developer like William Levitt could whack together 150 houses a day in the potato fields of central Long Island using a production system of specialized work crews and prefab components, until more than 17,000 nearly identical four room "Cape Cod" boxes stood in Levittown, as the agglomeration was named".[256]

This literature encompasses writings concerning the American experience with post-W.W.II prosperity and its impact on settlement patterns and suburban development in pursuit of the American Dream. The majority of examples of this literature do not so much proscribe the important elements that must be present in a livable community, than describe and offer explanations as to why certain urban and sub-urban configurations occurred during the post-war period.

A confluence of socioeconomic and political factors influenced the creation of the modern United States cities which have uneven degrees of livable neighborhood attributes. Many authors cite rising real personal income, the rising cost of good urban land, the mass use of the automobile, the availability of cheap residential land outside of cities, growth and zoning controls, and the American ideal of home ownership subsidized through home mortgage tax exemptions as factors in explaining the demand-pull conditions for the formation of suburban cities on the outskirts of many older, industrial cities.

These new suburbs based around individual car ownership, severed the pedestrian-scale locational links between residential housing, shopping and employment, making the automobile essential for transportation in these new communities. The trend toward the post-war suburb was the built expression of the national desire to live the modernist conception of the "American Dream"; owning your own home with a big yard, separated land uses, and mobility provided by the automobile and free roads.

Prewar Visions of the Ideal Community

> "[Le Corbusier] proposed apartment blocks-lower than the office buildings, about sixteen stories-organized in long slabs that each could house three thousand people. Instead of streets and sidewalks, there were elevators and corridors...It was all very rational—"Cartesian" was a favorite Le Corbusier term—cars over there, living over here, work above, play below. Voilà! The city of the future".[257]

While Le Corbusier could not envision the role of the suburb in the city of the future, what he called The Radiant City,[258] Americans, with the help of the automobile could. The separated uses envisioned by Le Corbusier influenced a generation of architects and planners who set about implementing this vision but with a few changes. The apartment blocks were typically built as low-income housing (urban renewal projects) and the suburbs were connected to urban skyscrapers by the interstate highway system.

As the US was still recovering from The Great Depression, the New York World's Fair in 1939 displayed an exhibit from General Motors called "Futurama". The exhibit was a look ahead in time where Americans would drive on super-highways to skyscraper cities and live in a low density, bucolic utopia. The message of the Fair was that universal automobile ownership and the highway would lead to the realization of the American Dream. It would provide high degrees of mobility and an escape from the social conflicts and poverty of The Great Depression.[259]

This vision was largely realized as automobile ownership increased after the war. Government support for home ownership through Federal Housing Administration and Veterans Administration housing loans, massive slum clearance through urban renewal, and later, the building of the interstate highway system reinforced this trend. America became a nation of single family homes on private lots connected by the automobile.

Building the Postwar Suburb

> "The suburban subdivision was unquestionably a successful product. For many, it was a vast improvement over what they were used to. The houses were spacious compared to city dwellings, and they contained modern conveniences".[260]

Many post war suburban developments and certainly most new suburban developments were not at all unplanned. In fact, many of these communities were stringently planned and controlled down to the color a resident could paint their house. This control and planning was largely meant to maintain the property values of current residents resulting in potential buyers facing higher prices than would otherwise be the case. The results were spatial separation of land uses connected by large arterial roads, regulated low density development, and expensive homes.[261]

World War II is frequently considered a turning point in the development of cities. Both during the war and as the war came to a close and soldiers were returning home, the US government made policy decisions that greatly affected urban form in the post war years, even to the present. To support the war effort, the vast military industrial complex developed in new, undeveloped areas across the country. Corporations like Hughes Aircraft, Lockheed, and General Dynamics built enormous factories that required a huge number of workers. Very large scale developments were built with large tracts of homes, similar to the suburban tract developments common today.[262]

These developments provided housing and a lifestyle to the war workers and an expectation about the future among soldiers returning home. Many, including President Truman, felt that those who contributed to the war should be rewarded with, among other things, a new single family house on a private lot if they desired.

This policy was reinforced by:

The mortgage interest deduction to subsidize home ownership,

Government subsidized, FHA and VA loans for home buyers,

Rapidly increasing levels of automobile ownership,

Direct government subsidies for roads, sewers, and highways, and

Popular visions of idealized suburban life.

Unintended Consequences

"In almost all communities designed since 1950, it is a practical impossibility to go about the ordinary business of living without a car. This at once disables children under the legal driving age, some elderly people, and those who cannot afford the several thousand dollars a year that it costs to keep a car..".[263]

As with many markets, external economic effects (i.e. "externalities") were generated by the low density development patterns promoted after the war. For example, driving an automobile causes pollution, yet the polluter does not compensate society for this cost. Economic efficiency requires that the behavior causing negative external impacts be curtailed in some way,[264] preferably through appropriate pricing policies.[265]

As low density development expanded to include increasing amounts of land area, the supply of undeveloped land decreased causing higher building costs. The spatial distribution of development across vast land areas required increasingly distant commutes and travel times. The results were large external impacts of increased congestion, pollution, and increasing transportation costs. All of these effects are considered negative externalities in traditional transportation economics.

These development patterns were codified and reinforced by transportation infrastructure investments, zoning and building regulations, and tax policies that promoted the over-consumption of land resources. Initially, people moved to low density developments to escape the perceived crowding, traffic, and pollution of cities. As more and more people moved to the new areas, these undesirable characteristics soon followed.

The literature is replete with references to the consequences associated with contemporary American demographics and living choices. Numerous essays and studies address the American life-style and its impact on travel demand,[266] the formation of low density suburbs,[267] and the general de-evolution of the urban city center. Starting in the 1970s, the United States government commissioned studies of the societal costs associated with American living preferences. For example, *The Costs of Urban Sprawl*,[268] examined the negative impact of the country's low-density settlement patterns on environmental, fiscal and energy grounds. While this official view did lend support to urban planners favoring high-density developments, (see next section) it did not address the value that some Americans place on low-density settlement patterns. Indeed, authors Audirac, Shermyen and Smith,[269] in defending Florida's growth management decisions, cite numerous studies that support a clear preference for low-density living.

While the government was studying the costs of urban sprawl, the subsidization of sprawl continued unabated. Government support for highways and infrastructure subsidies for suburban growth made transit not only an inferior choice for many citizens but an impossible choice for many in the suburbs. The low-density development patterns made possible by free, government provided roadway infrastructure made transit inconvenient in the suburbs. Coupled with the lack of capital funds for transit systems meant that many residents had no access to transit.[270]

Urban Renewal: Modern Urbanism Applied

> "The government-funded low income housing projects were built on the old existenzminimum principle of a Weimar siedlung, which assumed workers have no higher aspiration for the quality of life than to be stacked like anchovies in a concrete can".[271]

Urban renewal has been nearly universally derided as a failure of well intentioned policy. During the great depression and W.W.II, urban development in the United States was nearly non-existent. There was almost no downtown development, excessive unemployment (until the war) and factory closings. Even suburban housing development ground to a halt.

Most major cities had been starved for investment for nearly twenty years resulting in acute housing shortages and poor living conditions. Responding to the need for new housing and the governments desire to reward the country for sacrifices during the depression and war years, cities became the recipients of massive inflows of government spending. Cities benefited economically from the construction jobs but suffered intense social upheaval as entire neighborhoods were razed to make way for high-rise public building projects. Such projects destroyed neighborhoods, shopping areas and the social fabric of cities across the nation.[272]

Poor neighborhoods that were not razed often were cut up by multi-lane freeways. The interstate highway system, while providing a flood of short term jobs and improved suburban access to central business districts, created physical barriers between urban neighborhoods and districts. Further, the system accelerated urban neighborhood decay by allowing easy access to suburban residential areas.

The demolition and replacement of large sections of urban areas necessarily disturbed the life of cities and came at some human cost. However, the result was supposed to be an improvement upon the status quo. In the case of American urban renewal projects, the results were not impressive; in fact more housing was razed than was constructed over the life of urban renewal. The physical impact of urban renewal was driven by the notion that cities had to be modernized according to the emerging urban ideal,[273] known as modern urbanism, summarized by:

the primacy of the automobile

streets for transportation only

pedestrians relegated to separate walkways and pedestrian malls

separated land uses

high-rise buildings set in courtyards of open space

large buildings separated from other parts of the city

The largest application of these principles were the public housing projects of the 1950s, 1960s and 1970s. Large tracts of land were condemned by the government, cleared and set aside exclusively for public housing. The public housing was built as enormous high rise structures on so called "superblocks". Between the widely spaced buildings were public parks and all business, shopping and entertainment were removed from these public housing project areas. Presumably, residents would be required to drive to shops and to work.

Unfortunately, public housing residents often could not afford an automobile resulting in intense dislocation from the social fabric of the city. [*Urban unemployment caused by segregation of the poor away from shopping and employment areas has been hypothesized by John Kain. The theory is known as the spatial mismatch theory.*[274]]

This urban structure violated the principles that make cities work. In fact, the design of modern urbanism was essentially the design principles laid down by Le Corbusier nearly thirty years before. The acceptance of this urban development paradigm ensured that what seemed like a ridiculous idea before the war, now seemed sensible.[275]

The Urban Neighborhood: A Paradigm Shift?

> "As anyone who reads fiction in the New Yorker knows, Americans mostly live in banal places with the souls of shopping malls, affording nowhere to mingle except traffic jams, nowhere to walk except the health club".[276]

Unprecedented urban fringe growth was made possible by the automobile and government financed freeway building. This coupled with the emerging evidence that urban renewal programs in the 1950s and 1960s were failing to live up to their intended purpose, led to growing criticism of the emerging urban structure.

In city centers, the renewal effort had applied suburban style development principles to high density urban settings with unimpressive results. Large mixed-use neighborhoods were bulldozed and converted to massive housing developments separated from commercial and industrial areas of the city. Freeways were built to connect suburbs to the city core, further isolating residents closer to the city center. Kain and others suggested that the result was a spatial dislocation among urban residents, removing access to shopping, employment, education and social services for large numbers of (predominantly poor) people.

High levels of traffic congestion characterize many of the largest US cities. Congestion, expensive public services, dwindling tax base and social dislocation in center cities became obvious problems that demanded a response.

These problems paved the way for emerging ideas regarding livability. These ideas were in response to the rapid, suburban tract oriented growth of the post war period, inefficient levels of traffic congestion and the failures of the urban renewal movement. Urban philosophers, such as Jane Jacobs,[277] William Whyte,[278] and Lewis Mumford[279] supported the vision of an urban, moderate to high-density, mixed land-use city in which residents had easy access to employment, shopping, and leisure activities via walking and well-developed mass transit systems. This vision contrasts with suburban tract development and urban renewal, which attempted to separate land uses and connect various uses by an automobile–oriented transportation network.

The livable community idea is not, however, only a central city issue. Many livable communities and most potential transit-oriented developments will be located in the suburbs where the development of new types of communities is still possible. The changing socio-economic, demographic and political landscape of the last decade has contributed to the recent literature favoring the creation of urban, livable neighborhoods.

In addition, many suburban townships and city governments, are now faced with the dilemma of either luring more retail shopping malls and outlets or raising local property taxes to provide the necessary revenues to maintain adequate public services.

Livable Neighborhoods in Practice

Realizing that current development patterns may be causing many urban neighborhoods to become less "livable", policy makers have begun looking to operational methods for improving the livability of neighborhoods.

After building cities around the presupposition of the automobile for so long, it takes a concerted effort to identify ways to ameliorate the negative effects of an automobile based society and to promote communities that are transit accessible and livable. One of the first references to the term, "livable community" is in a planning study conducted in Vancouver, British Columbia. Several cities in the United States have followed, notably, Portland, Oregon. These efforts are detailed below.

Vancouver, British Columbia--False Creek Development The plan for redeveloping the False Creek area near downtown Vancouver began in the early 1970s. The goal of the project is to involve public responses in defining "livability" issues and how the area can be redeveloped to serve the needs of residents and to maximize the "livability" of the region.[280] The issues addressed include:

Making streets more inviting to pedestrians

Ensuring usable public space

Enhancing access to the waterfront

Integrating the area into the regional transit service and improving access

Determining the optimal residential land use density

These issues deal with urban design as it relates to the desire for a comfortable, convenient, and affordable place to live in an urban setting. The resulting plan highlights the importance of community involvement in developing livable environments.

Portland, Oregon--Livable Cities Project Supporting the development of livable communities has become a priority in several areas around the United States For example, the *Oregon Transportation Plan* (OTP) Policy Element addresses the issue of livability under its Goal 2: "To develop a multimodal transportation system that provides access to the entire state, supports acknowledged comprehensive land use plans, is sensitive to regional differences, and supports livability in urban and rural areas".[281] These goals are divided into four topics: land use, urban accessibility, rural accessibility, and aesthetic values.

Decentralized development has "tended to separate residential areas from employment and commercial centers requiring people to drive almost everywhere they go. The result has been increased congestion, air pollution and sprawl in the metropolitan areas and diminished livability".[282] To accommodate population growth and protect livability, Oregon will adopt land use policy as a primary development tool.

Transportation systems development will need to support concepts of mixed use land development, compact cities, and connections among various transportation modes to make walking, bicycling and the use of public transit easier. In turn, land use plans and development need to support the policies and objectives of the transportation system plans.

The Transportation Planning Rule (660-12) prepared by the Oregon Department of Transportation (ODOT) and the Land Conservation and Development Commission (LCDC) encourages reduced use of the automobile and requires planning for the use of alternative modes of transportation in urban areas.

Literature promoting the urban livable community ideal is a staple of the planning community, but it is also supported by environmental and public policy practitioners. Environmental and conservation concerns, such as the Conservation Foundation,[283] support the ideal of livable communities that emphasize sustainable growth policies in resource management, including historic and cultural properties and districts, aesthetic resources and open spaces. Public policy forums now stress the need for cooperative efforts between private and public interests to develop new urban forms to counteract the forces of urban sprawl.[284]

Seaside, Florida Development The small resort town of Seaside challenges the dominant development paradigm with its neo-traditional design principles. The inspiration for this community is really the streetcar suburb of the late 19th and early 20th century. Seaside was planned and designed by Andres Duany and his wife and partner, Elizabeth Plater-Zyberk; two of the early promoters and best-known architects of the "new urbanism". The model for Seaside is not the high density city, but the moderate density "village".

Seaside is distinguished by narrow streets in a walkable grid, and a jumble of pastel colored homes with mandatory front porches. It has a town center with an elliptical commercial district and a grand boulevard. The town center includes many apartments that provide affordable housing for the people who work in the commercial district. Single family houses have "granny" units over the garages to promote income mix in the neighborhood.

Seaside has been praised broadly as an example of how the application of traditional town planning principles can create good relationships between public and private space and simultaneously lead to a commercially successful venture.

One development lesson that emerges from the Seaside experiment is the role of building codes and planning regulation. When Seaside was planned and built, Walton County had no building codes and no official inspector. The developer and the architects were free to make their own rules and impose their own code upon future development in Seaside. Most development codes, Duany and Plater-Zyberk note, are impediments to traditional town planning and create incentives for continued low density sprawl.

The missing link in these new neo-traditional towns is transit. The reason is not for lack of demand or design considerations, but a lack of investment in transit systems to serve new communities since most transit systems in the US have been built to serve existing neighborhoods.

In the past, these types of towns depended on and reinforced use of the transit system. Only new transit starts and public effort to integrate these new towns into urban areas will show whether the features of these towns can strengthen transit use and be strengthened by transit access.

Criticism of Building Livable Spaces

"The idea has sold more books than houses".[285]

Since the conception of livable communities, urban theorists and journalists have been the only sources of much of the early work expressing the value of "livability". Their approach is sometimes criticized for being self serving. Audirac (1990) and her co-authors assert that many urban planners neglect the preferences of the vast majority of Americans who prefer to live in distinctly "unlivable" (i.e. auto dominated, low density, residential use only...etc.) communities.[286] These preferences are reflected in the revealed preference of the majority of home buyers. To critics, the market provides better information about preferences than planning professionals.

The results of Audirac's research, while showing a distinct preference for low density living, also shows that this preference is not universal. She notes that the most preferred locations in Florida are downtowns and rural or semi-rural locations, with suburban locations the least preferred. This finding implies that there is a wide diversity of opinion about preferred living environments.

Many critiques of livable community initiatives point out that most surveys of individual preferences indicate that a majority of people desire to live in low density, automobile dependent environs. The study by Audirac, et al (1990) cites a litany of disbenefits to density and proximity to services.

The authors find that proximity to non-residential properties such as pharmacies and markets actually decrease neighborhood satisfaction.[287] Many studies have found that density is correlated with a dislike of interaction with dissimilar groups.[288] Density is found to interfere with residents ability to regulate contacts outside their homes and create perceptions of crowding and loss of social control.

Studies that challenge the academic and urban planning views on individual preferences for livable community attributes point to some conclusions about the measurement of their values. The authors point out that what urban experts have defined as a livable community has changed over time and does not lead to a consistent vision of preferred community attributes. The quantitative research conducted for this study will shed light on the question of whether residents are willing to pay a premium for transit-oriented neighborhoods.

Transportation Policies and Neighborhood Development

Transportation is integrally related to land use and, by extension, to the form of the neighborhoods that private developers build and that residents demand. Land use development is driven by the balance between the costs and benefits of various development patterns. A major determinant of the costs of living in a given neighborhood is the cost of accessing work, stores, friends and family, and other institutions outside the household. The transportation infrastructure, as a primary determinant of access, affects the costs of different types of neighborhood developments in a variety of ways.

Transportation and land use are inter-related. Transportation systems influence the relative prices of various development patterns influencing the land use mix while development patterns influence which type of transportation infrastructure is demanded. But if one influence has to be chosen as the prime mover, analysts increasingly choose transportation. The most significant changes in land use development over the past 150 years have resulted from changing access patterns provided by major changes in transportation.[289]

The importance of the result that transportation strongly influences development patterns lies in the fact that public policy plays a major role in determining which transportation infrastructure investments occur and what the transportation system will eventually look like. Land use on the other hand is largely driven by market forces, regardless of such government policies as zoning, given the characteristics of the transportation system. Through transportation policy, government can influence perhaps the most important determinant of urban form.

Transportation Improvements Affect Land Use Practices

No model accurately predicts land-use and travel patterns in a polycentric city. Most models simply predict short run impacts on travel patterns from transportation system improvements, but tend to fall short of predicting long term changes in travel behavior. Urban structure (land use intensity for example) has an impact on demand for transportation services. Long term changes in urban structure, brought about by changes in the transportation system, remain unquantified in the current analysis of transportation projects, which may bias estimates of travel behavior and benefits when structural changes occur.

Transportation and Land Use Impacts To illustrate the possibility of significant effects from long term structural change in urban form, consider a transit project that stimulates increased urban densities around the transit stations over a period of many years. Current demand estimates are based on ridership surveys which ask potential riders of their willingness to pay for transit services. If the project causes structural change in urban form, these ridership surveys will produce short term demand elasticity estimates that are different from the long term demand elasticity.

Several studies have found a correlation between land use intensity and daily vehicle miles traveled (VMT). A recent analysis has estimated that VMT may be as much as twenty-five percent lower in transit-oriented developments as opposed to traditional low density suburban subdivisions.[290] If investments in transit projects increase land use intensity around transit stations in the long term, then those who live near the stations will alter their travel patterns in ways that can be accounted for in standard transportation demand models.

As an example, consider a project expanding a transit system. This increased capacity of the transportation network may allow firms to consolidate their office operations at more central locations (in the same manner as manufacturing firms consolidating their logistics operations as a result of expanded road network capacity).

The effects of long run restructuring suggests that the long run impacts of transportation investments on land use are not quantified when infrastructure investments are made.[291] Figure 5.3 presents the short, medium, and long term impacts of transportation investments.

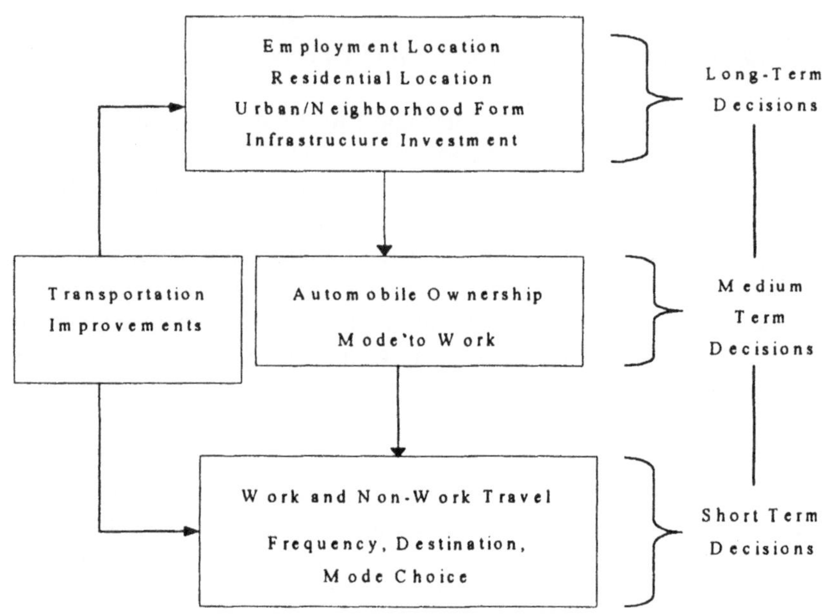

Figure 5.3 Transportation Improvements Affect Land-Use

Source: Moore, Terry and Paul Thorsnes, *The Transportation/Land Use Connection: A Framework for Practical Policy*, (Washington, DC: American Planning Association 1994).

There is little question that the long run impacts of transportation investments have influenced the way we build cities and neighborhoods. Transportation planners are well aware of cases where highway improvements projected to accommodate fifteen years of traffic growth are choked with congestion in far less time.[292] One reason for this phenomenon is that the transportation improvement caused changes in land use over the long run that increased automobile use on the highway beyond what would have occurred without the investment. Unfortunately, land use patterns induced by expenditures on freeways reflect the problematic urban form which drives the current reaction against sprawling tract developments and toward the new urbanism.

Transit and Neighborhood Design The public policy issues involving the emerging neighborhood development paradigms revolve around transportation infrastructure. Transit access reinforces the new paradigm while new freeway developments run counter to it. This review assesses the potential benefits of new development paradigms such as TOD and the new urbanism and by extension, the neighborhood livability benefits of offering transit access in neighborhoods.

Transit access, by changing the relative prices of traveling to various destinations influences land use patterns. The possible effects of transit developments include increased densities and improved walkability. Many authors suggest that building transit systems do not automatically bring about major land use changes by itself. Due to the complexity of land use, other development incentives and policies are likely required to effectively promote the planner's conception of "livable" neighborhoods.

The effects of transit on land use will be strongest with the highest capacity and performance transit modes. Heavy (rapid) rail transit stations have the largest sphere of influence for land use, often assumed to be about a 0.5 to 1.0 mile radius from the station. Light rail transit stations influence, maybe, a 0.25 to 0.5 mile radius, and a bus station will probably have no land use impact at all.[293] Indicators of transit impacts on neighborhood livability will be of two predominant types:

Reductions in transportation expenditures

Changes in property values

These two indicators of transit's impact on neighborhood livability will be theoretically and quantitatively assessed below.

Transit and other transportation infrastructure are only a part of the determinants of livability in neighborhoods and the urban structure that develops around that infrastructure. Figure 5.4 expresses the multi-faceted and complex inter-relationships that help influence urban form and neighborhood livability.

Neighborhood Benefits and Development Patterns The average household spends nearly 20 percent of their total budget on transportation,[294] excluding time lost to congestion and long commutes, taxes to support the current transportation infrastructure, and environmental damage. This situation is perpetuated and codified in zoning laws and city development plans that typically disallow mixed-use development, small streets, multifamily units, small lot sizes, and "in-law" rental units.

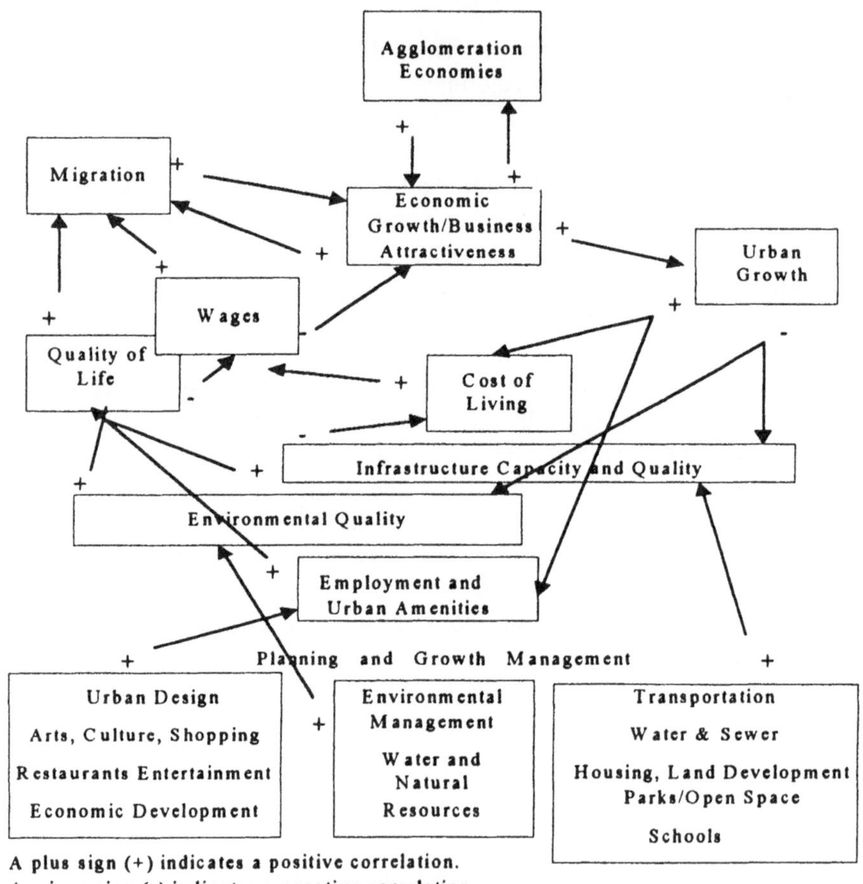

Figure 5.4 Linking Transportation, Urban Form, and Livability

Source: Dowell Myers, "The Ecology of 'Quality of Life' and Urban Growth", *Understanding Growth Management: Critical Issues and a Research Agenda,* (Washington, DC: The Urban Land Institute, 1989).

The policies that complement TOD will do the opposite. Streets are built small and walkable, and zoning deregulated to encourage the most economically productive use of scarce land resources. Infill of unused space is encouraged to increase density and prevent the destruction of open spaces in undeveloped areas. However, suburban infill often faces the opposition of neighborhood groups that want to inhibit density and mixed-uses while limiting growth in their area. The predictable result is higher land and development costs for the entire region.[295]

Portland, Oregon is an example of a region that supports infill using a region-wide urban growth boundary to prevent urban sprawl outside a specified urban region. Portland area governments have also implemented zoning supportive of transit-oriented development around light rail stations. This public policy support for TOD helps to identify Portland as an ideal case study community for measuring any potential benefits from TOD.

Travel Behavior and Transit-Oriented Development

If future development patterns perpetuate current automobile-oriented development (AOD), public policies aimed at reducing auto use are destined to be less effective than policy makers hope. Congestion pricing, user fees, gas taxes, and other demand management methods will be painful with limited reorientation of travel behavior in the short run. The reason is that current development patterns make traveling by any mode besides the automobile a potentially difficult and time consuming prospect. Transit will be inconvenient unless people can walk or bike to transit stops and can comfortably walk to their final destination after riding on the transit system.[296]

The implication is that public transit investments, in the absence of complementary TOD policies will result in under-utilized transit facilities. Since development patterns can influence the way in which people travel, TOD will likely have a positive feedback effect into transit usage while AOD has a positive feedback effect on automobile usage.

The following statistics reflect national transportation increases for the period 1969 to 1990:

Population	21 percent
Trips per Household	50 percent
Vehicle Miles Traveled	82 percent

The figures illustrate the effects of increasingly fragmented development patterns. The more people move to new AOD's, the more they are required to travel by car. A recent study by John Holtzclaw (1991) found that density and vehicle miles traveled (VMT) are significantly linked with higher densities reducing VMT.[297] However, early studies of VMT and density failed to account for variables like income, employment and household type that may be the real predictors of VMT. The relationship between urban design and VMT has been the focus of a great deal of recent research. John Holtzclaw modeled the density, transit access, VMT relationship accounting for income effects and found that higher density and better transit reduce VMT. In fact, Holtzclaw found that income is not a significant predictor of VMT when density and transit access are included in the model.[298]

There is some evidence from travel surveys that the type of neighborhood that one lives in helps determine travel behavior. A study by Fehr & Peers Associates for the International Association of Traffic Engineers compared the travel behavior residents in older, traditional small towns near San Francisco with residents of new suburban tract developments. Using travel surveys conducted in 1980, they generated the results given in Table 5.1.

Table 5.1 Travel by Neighborhood Type

	Auto Oriented Suburb	Traditional Neighborhood
Daily Trips per Household	11	9
Modes	Percent	
Auto Trips	86	64
Transit Trips	8	17
Walk Trips	3	17
Bike Trips	3	2

Source: Fehr & Peers Associates

The difference in travel behavior is clear though the real reason for the difference is not. A possible reason can be that the structure of these particular types of communities influence how people travel, or people may simply be settling in areas that reinforce their lifestyle preferences. People who buy homes in an AOD may simply like to drive more than other people. The results may vary if differences in income, accessibility to transit, and proximity to services and jobs were accounted for in the analysis.

Evidence of VMT and Development Form John Holtzclaw, in his work for the Natural Resources Defense Council, has developed several mathematical relationships which predict VMT. He finds that the most important predictors of VMT are residential densities and an index of transit service. Holtzclaw tests the hypothesis that residents drive less when they live in neighborhoods with higher densities, more transit service, accessible shopping, and an attractive, pedestrian friendly environment.

If the relationship between transit access and neighborhood form and transportation costs can be determined, the cost savings from the improvement of transit service can be directly calculated. One measure of the benefits derived by residents of Transit-Oriented Developments may be the transportation costs savings experienced by residents compared to residents of more automobile dominated neighborhoods.

Americans spent between $775 billion and $930 billion on direct annual auto costs in 1990.[299] The average yearly expenditure per household approaches $10,000, or over $800 per month. This level of expenditure varies systematically between neighborhoods within urban areas and between urban areas based on several identifiable characteristics. Holtzclaw notes that denser, central areas with good public transit access, nearby shopping and employment require, not surprisingly, much less driving than sprawling, low density neighborhoods in the exurbs (see Table 5.2).

The difference between Danville-San Ramon and Walnut Creek, two suburban communities near San Francisco, is particularly interesting because both communities were low density bedroom communities prior to BART opening in the 1970s. Since two BART stations opened serving the Walnut Creek area, it has developed to over twice the density of Danville-San Ramon with nearly 4 times the local jobs. While BART cannot be ascribed all of the credit for increasing densities and job growth, the anecdotal evidence is strong. The lower auto travel per household in Walnut Creek allows households to spend $8,000 less on automobile costs compared to households in Danville-San Ramon who lack significant transit service.

Holtzclaw Curves The mathematical models used by Holtzclaw to predict auto ownership and VMT find that neighborhood density and a transit accessibility index are the best predictors of variation in auto ownership and VMT between neighborhoods.

Income variations are accounted for in the analysis and found to explain only limited variations; a result contrary to conventional wisdom. Holtzclaw suggests that the lack of strong income effects results from the choice of communities, none of which are particularly low income.

Table 5.2 Auto Expenditures in San Francisco Area Communities

San Francisco Metro Area Neighborhood	Auto Expenses per Household
Danville-San Ramon (Very Low Density)	$17,800
Walnut Creek (Low Density)	$9,800
Rock Ridge (Moderate Density)	$8,800
San Francisco (Moderate-High Density)	$6,800
Nob Hill—Fishermans Wharf (High Density)	$4,200

Source: John Holtzclaw, "Using Residential Patterns and Transit to Decrease Auto Dependence and Costs", Natural Resources Defense Council, 1994, p. 6.

Auto ownership decreases as density increases. This result is consistent with theory and with previous research. There are a number of reasons for density to reduce auto ownership.

Densely developed communities often have good transit service allowing residents to substitute transit trips for auto trips.

Increasing densities typically result from high land values where people economize space by living in closer proximity.

High land values often translate into high auto parking costs.

Savings in auto expenditures in denser, transit served neighborhoods do not accrue to residents without some offsetting expenses. In general, we expect neighborhoods which enjoy lower transportation costs to have higher property values and rents than other neighborhoods. Economic theory predicts that, assuming consumers have perfect information regarding costs, two properties differentiated only by transportation costs will differ in value by exactly the amount of those costs fully capitalized into property values.

Auto Ownership The relationship that best predicts auto ownership is found by constructing a series of regressions and testing alternative specifications. The explanatory variables in the analysis are density (in terms of households per acre), a transit accessibility index (number of 50 seat transit vehicles per hour), a neighborhood shopping index, and a pedestrian accessibility index. The best simple fit is found to include only one variable, density and explains 85 percent of the variation in auto ownership between California neighborhoods:

log (Autos per Household) = log (2.704)–.25 * log (density) + error term

$R^2 = 0.850$

This relationship for California is shown in Figure 5.5. It is expected that the shape of this curve should hold for most regions although the actual values will be region specific.

Annual Vehicle Miles Traveled Holtzclaw confirms that VMT per household increases as residential density, transit accessibility, neighborhood shopping indices, and pedestrian accessibility indices decrease. The best fit regression equation explaining 83 percent of the variation in VMT between neighborhoods incorporates density and the Transit Accessibility Index (TAI)[300] in the following form:

log (VMT per Household) = log (34,270)–.25 * log (density)–.076 * log (TAI) + error term

$R^2 = 0.830$

This relationship is shown in Figure 5.6. The four lines represent different levels of the transit accessibility index while density and VMT reside on the axes. As expected, the relationship between VMT and density is similar to the auto ownership curve and improving levels of transit service tend to reduce VMT at any density.

Auto Costs per Household By combining these functional relationships between VMT, auto ownership, density and transit service with estimates of the average cost of auto ownership ($2,203 per auto annually and $0.127 per mile–Federal Highway Administration, 1991), Holtzclaw develops a functional relationship for predicting annual household expenditures. The equation is as follows:

Annual Household Auto Costs = $\$2{,}203 * 2.704(\text{density})^{-.25} + \$0.127 * 34{,}270(\text{density})^{-.25} * \text{TAI}^{-.076}$

This relationship is simply the sum of the auto ownership and VMT equations parameterized by unit costs to derive an estimate of annual auto costs. Figure 5.7 presents the predicted annual household expenditures on direct automobile costs as derived from Holtzclaw's California data.

Vehicles Owned per Household

Figure 5.5 Auto Ownership and Residential Density

Source: John Holtzclaw, "Using Residential Patterns and Transit to Decrease Auto Dependence and Costs", Natural Resources Defense Council, 1994.

The relationships developed here should be generalizable to derive a model that calculates auto ownership, VMT and automobile expenditures per household given any input of density and transit service characteristics. The change in expenditures can then be calculated for varying levels of transit service and density levels.

The Holtzclaw study uses cross sectional data that relies on the variation in densities and transit service across communities at a single moment in time rather than looking at communities as transit service and densities change over time.

Application of the 'Holtzclaw Effect' Research by John Holtzclaw[301] can be applied in a straightforward way, to estimate the effects of various types of communities on automobile expenses.Communities with varying characteristics can then be compared to determine cost savings from better transit service and per or higher residential densities.

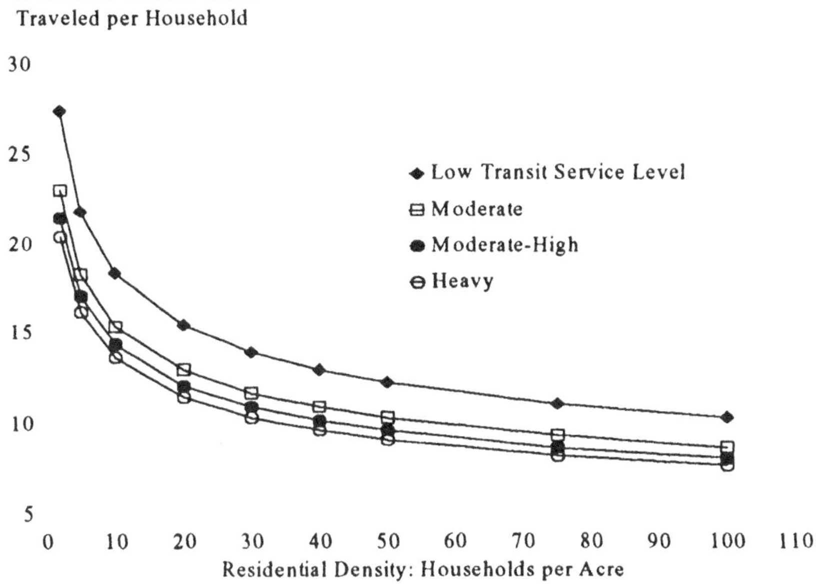

Figure 5.6 Predicting Auto Travel

Source: John Holtzclaw, "Using Residential Patterns and Transit to Decrease Auto Dependence and Costs", Natural Resources Defense Council, 1994.

Residents who enjoy better transit service are less likely to own a car and drive the cars they own less frequently. Predicted auto expenses do not show as much variance in response to transit service as VMT because many people who use transit to reduce VMT still incur the costs of owning an automobile.

Comparing Auto Expenses in Development Scenarios

The model results presented in the graphs above can be used to compare the expected automobile expenses in various development and transportation scenarios. The following tables present a series of scenarios in which density and transit service are varied and the household auto expenses calculated.

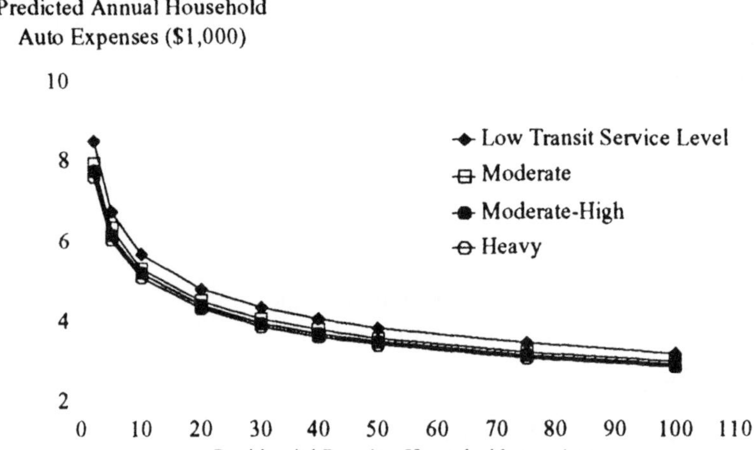

Figure 5.7 Predicting Automobile Expenses

Source: John Holtzclaw, "Using Residential Patterns and Transit to Decrease Auto Dependence and Costs", Natural Resources Defense Council, 1994.

Transit Service Variability Scenario The first scenario (Table 5.3) compares two neighborhoods with the same residential density where transit service improves from 10 transit vehicles an hour to 80 transit vehicles an hour. This scenario approximates the difference between frequent bus service and a rapid rail transit station with a bus station. Aggregate savings are calculated for residents within a ½ mile radius of the transit stop.

The results of this scenario analysis show that average annual auto costs for residents of the high transit service area were $300.00 less per household than for residents who enjoy lower transit service. Cutting transit service from a level of 80 fifty seat transit vehicles to 10 will cost the average resident over $25 per month in additional automobile expenses. Assuming density of 10 households per acre is constant across a given metropolitan area with 75 rapid rail transit stations with service levels approaching 80 vehicles per hour, reducing transit service to an index of 10 would result in added annual auto expenses of about $113 million for residents living near those stations; about 380,000 people.

Table 5.3 Auto Expenses Saved from Transit Planning

Calculated Auto Cost Savings in Transit Oriented Developments

	Annual Auto Costs	Household Density	Transit Service Index
Auto Oriented Scenario	$5,404.37	10	10
Transit Oriented Scenario	$5,104.03	10	80
Yearly Savings from TOD	$300.34		
Monthly Savings	$25.03		

Total Expenditure Effects in ½ Mile Impact Area

	Auto Oriented	TOD	Acres
Households–½ mile	5027	5027	502.65
Total Yearly Savings	$1,509,657		
Total Monthly Savings	$125,805		

Alternative Neighborhood Development Scenario The second scenario (Table 5.4) compares the auto expenditure impacts of two development scenarios. The first is the typical auto oriented development while the second typifies a "neighborhood" transit oriented development. Auto oriented suburbs have a typical household density of 2.5 household per acre compared to 10 in a transit oriented development. Transit service in auto oriented suburbs can be infrequent or nonexistent. In this scenario, we assume 2 buses per hour serve the neighborhood while the transit oriented neighborhood is served by rapid rail and frequent bus service.

The results of this scenario show a strong savings in automobile expenses. Average annual auto savings in the higher density per high transit access neighborhood was $2,917 per household. The result shows how the development patterns of neighborhoods can either reinforce or inhibit certain travel behaviors. The first scenario showed that adding transit while maintaining development patterns provides significant savings in auto costs, but by allowing density to expand from 2.5 households per acre to a still modest 10 per acre, the savings per household increase over 10 times the first scenario. The household automobile expense savings from living in one transit oriented neighborhood exceed $14.5 million.

Transit's Impact on Urban Form

In the Holtzclaw analysis, urban form dominates the transportation mode choice. However, transportation infrastructure and urban form are interrelated so that over the long run (allowing the urban form to change in response to the transportation system), the VMT and auto cost impacts of transit service may be much stronger.

Table 5.4 Traditional Suburb with Minimal Bus Service (2 Buses per Hour) Compared to a Residential TOD Served by Rapid Rail Transit Line and Bus Hub

Calculated Auto Cost Savings in Transit Oriented Developments

	Annual Auto Costs	Household Density	Transit Service Index
Auto Oriented Scenario	$8,020.99	2.5	2
Transit Oriented Scenario	$5,104.03	10	80
Yearly Savings from TOD	$2,916.96		
Monthly Savings	$243.08		

Total Expenditure Effects in ½ Mile Impact Area

	Auto Oriented	TOD	Acres
Households–½ mile	1257	5027	502.65
Total Yearly Savings	$14,662,083		
Total Monthly Savings	$1,221,840		

While the relationship between transportation infrastructure and land use is accepted generally, quantifying the impact of transit investments on station area density is controversial and difficult.

Several studies have been conducted to identify the relationship between transit station proximity and the intensity of development, with mixed degrees of success. A study analyzing the Washington, DC Metro system[302] finds a significant link between station proximity and development intensity. The authors conduct three separate modeling exercises. The first compares the Metro station areas and rail corridors with control areas to determine whether transit accessible areas experienced higher population and employment growth rates than the controls. The second study compares rail corridors to other transportation corridors without rail transit service. The third approach divides the rail corridors into station areas and non-station areas and examine the growth rates. The results of each study confirm that station areas and rail transit corridors grow faster than areas without this accessibility.

Most studies of the link between rail transit stations and development intensity imply that while station areas confer advantages to properties directly accessible to the stations, real estate development is complex and is driven by a large number of factors, with transit access as one.

Measuring the Value of Transit-Oriented Development Attributes

The literature encompasses both purely qualitative and quantitative techniques for estimating or expressing the value of livable community attributes. Qualitative and quantitative methods are not mutually exclusive. *Qualitative* methods include essays by urban theorists (e.g. Jane Jacobs, William Whyte, Lewis Mumford[303]) and journalistic work (e.g. Neil Pierce). Other qualitative methods include focus groups and survey techniques although the results of these methods can also be analyzed quantitatively. *Quantitative* methods include modeling and simulation techniques. Quantitative methods often build on qualitative research in an effort to give qualitative descriptions and expressions a numeric value.[304] The methods that lend themselves most readily to the analysis of the benefits of transit oriented neighborhoods are stated preference methods and hedonic (property value based) models.

Qualitative Measures for the Value of Transit

Qualitative methods refer to attempts to express what people value about cities and livable communities. This approach emphasizes that attributes of livable communities are difficult to identify and measure and may be most usefully expressed in terms of people's belief systems and value judgments.

Livable communities are often assumed in the literature to be valuable by their nature, with no attempt to assign values to the attributes. Authors using qualitative approaches will value livable community attributes based, to some degree, on their belief systems and values about what a community should look like and how they would like to live themselves.

Neil Pierce (1993), among others emphasizes the need for empowerment in developing and maintaining successful communities. This qualitative approach seeks to identify attributes of communities that are prerequisites for the formation of livable communities. Pierce argues that empowerment of citizens is the key to creating livable communities. Pierce sees livable communities as neighborhoods "with a genuine sense of community, a place with an internal support system that functions family to family, neighbor to neighbor, the kind of neighborhood that many of us can remember as a vibrant, caring place".[305]

Empowerment is both a vehicle for creating livable communities and a goal in itself. The value of empowerment, Pierce says, is evidenced by well maintained and pleasant environments, safe and secure neighborhoods, and economic growth.[306] The idea of empowerment is expressed in the Federal Transit Administration, *Livable Communities Initiative*, as promoting increased participation by neighborhood and community organizations in the transportation and community planning process.[307] The value of empowerment resides in that it is a necessary, but not sufficient condition for creating livable communities.

Quantitive Measures for the Value of Transit. Quantitative methods for valuing livable community attributes usually build on qualitative approaches and attempt to rank or value communities based on their attributes. The methods employed in this can be grouped into the following broad categories:

Hedonic Wage and Price Estimation

Stated Preference Methods

Hedonic Methods Hedonic methods attempt to estimate a price for a public good by looking for a surrogate market. The surrogate market approach looks for functioning markets for goods and services where specific attributes (public goods) will be capitalized into the value of the observed goods or services. This surrogate market is observed where the attributes are deemed to be present and where they are deemed to be absent. Assuming perfectly functioning markets and market clearing prices, the value of attributes will equal the difference between the observed prices in the two markets. Hedonic methods are most commonly used in valuing environmental benefits or disbenefits (Freeman, 1993). Since livable community attributes can be easily thought of as positive environmental or neighborhood attributes, the extension of hedonic methods to valuing livable community attributes is theoretically straightforward.

Hedonic price estimation is performed using multiple regression techniques where the change in property values or wages are a function of community amenities (livable community attributes) and other social and economic variables. The regression coefficients are then used to calculate the implicit *marginal* prices of the amenities.

Stated Preference Methods Preference survey methods are a bridge between qualitative expressions of attribute values and quantitative methods for assigning economic value to those attributes. Preference surveys are typically based on consumer utility theory which states that people will choose a set of attributes that maximizes their utility.

The contingent valuation methodology ("stated preference" in the transportation field) is a technique for evaluating preferences, estimating utility functions, and forecasting demand for certain attributes or choice sets that do not have explicitly observable market mechanisms.

A typical application of the stated preference methodology is estimating travel demand functions.[308] The main reasons for using stated preference rather than revealed preference are:

It may be difficult to obtain sufficient variation in revealed preference data to examine all variables relevant to the model;

Explanatory variables in revealed preference studies may be strongly correlated; and

Revealed preference cannot be used to evaluate demand under conditions that do not yet exist.

A contingent valuation experiment is typically comprised of a list of choices or choice sets that are then administered to consumers in a survey format. The respondent states their preferences directly rather than revealing their preferences indirectly. As with any method that requires survey techniques, the validity of the results will depend critically on the format and design of the questionnaire.

The current study weighed the benefits of stated preference and revealed preference (hedonic study) approaches and chose to focus on hedonic methods. The primary reason is that the methodology is well established and offers a base on which we can build with new and innovative studies and because the need for using survey instruments make stated preference studies much more expensive. Future approaches to this topic could usefully apply the stated preference methodology to assess the livability of transit-oriented neighborhoods. We leave this methodology for future studies.

Results from Previous Studies Over the last several decades, authors have explored the impact of rail transit on property values. In general, research has confirmed that transit access usually provides benefits to residents that are capitalized into property values. The hedonic methodology was applied in most of these studies, providing a body of evidence and an established methodology for assessing the value of neighborhood and transportation characteristics. The following sections summarize the results of these studies.

Hedonic Studies

Washington Washington Metrorail (serving the Washington, D.C. Metropolitan area) is comprised of 83 stations on a 101 mile network. A number of studies have estimated the effect of these station on the surrounding property values. In one such study by Gatzlaff and Smith (1993), it was found that the average price for a townhouse within 1,000 feet of the station was $12,300 higher than comparable units further away. Lerman et al. (1978) found that for a single family home, a 10 percent change in distance results in a 1.3 percent change in property value.[309]

It is clear that the greater the distance between property and a transit station, the smaller the impact of the resulting property value effects. This relationship becomes stronger when commercial property values are examined. Lerman et. al. (1978) observed that retail properties are highly sensitive to transit proximity. A 10 percent change in distance from the station resulted in a 6.8 percent change in retail property values.

Another study found that commercial projects next to station areas in Bethesda and Ballston, demanded a $2.00 to $4.00 per square foot rent premium than similar projects a few blocks away.[310] We can assume that this is a result of a greater availability of both labor and customers.

The data presented above illustrates the sensitivity to station location that retail property experiences. In Washington "...even in corridors where development was slowing or declining, station areas still seem to be (relative) centers of economic activity and growth".[311]

Virginia is an area in which new development is occurring in proximity of Metrorail. According to a study by KPMG, it is estimated that the existence of the Metrorail will generate: commercial development of 26.8 million square feet; permanent employment of 91,000; and net tax revenues of $1.2 billion.[312]

Philadelphia The most examined rail line in Philadelphia is the Lindenwold, a 14.5 mile, 13 station line running to Philadelphia through the New Jersey suburbs.[313] This line has been the subject of intense study over the years, providing a rich repository of data showing the impact of transit stations on property values. In an early study by Rice Center, effects of station location on property value are reported to be about a 7 percent premium, or $4,500 per house.[314] Another study (Voith, 1993) observed that areas with commuter rail service enjoy house value premiums of $5,594...6.4 percent of the 1980 median house value of $87,455.[315]

The study by Voith went on to describe the neighborhood of the station area to determine the full impact of benefits per costs generated by station location. It was hypothesized that residential locations near rail service to the Central Business District (CBD) would have significant house premiums, more residents who work downtown, and lower auto ownership rates.

The station areas studied contained 29 percent more CBD employees than non-station areas. Voith expressed that the "productivity of the CBD and the transportation system are not independent, as one of the major attributes of the CBD is its accessibility to a wide labor pool". Rail transit facilitates that accessibility and is supported by the fact that the "...average car ownership in the sample is 1.63 cars per household...[4.5 percent] lower in [areas] with stations than in [areas] without stations".[316]

Rail stations provide accessibility to employment, as well as creating value added externalities, partly captured in the form of property values. Voith states that "over 40 percent of the residents of the suburban metropolitan area have a direct interest in the quality of public transportation and economic health of the CBD, regardless of whether they use the service or work in the CBD".[317]

Boston Boston is a city with a well developed transit system and numerous transit-oriented neighborhoods. Its first light rail transit system began operation in 1897 extending 28.5 miles with a daily ridership of 60,000.[318] Today, an extension of 7 miles is planned to encompass a dense residential area of 55 units per acre providing access to the CBD. The extension is expected to reinforce an already high-density, transit-oriented land use pattern.[319]

A study undertaken by R.J. Armstrong examines the Fitchburg per Gardner Line in Boston to quantify the neighborhood value created by commuter rail station location, captured in single-family residential property values. The study area encompasses 117,602 single-family residences. He found that property values in proximity of existing rail stations experience a 6.7 percent premium compared to property without rail access.[320] However, it is not by location alone that property values are affected. Travel time to and from the CBD contributes significantly to property values and is consistent with other studies.[321] Armstrong notes that for every 10 percent increase in commuter rail travel time, single family residential property values decrease by 13.7 percent.[322]

Exploring the micro effects, i.e. the immediate station location area, Armstrong discovered inconclusive results regarding property values. This is due to the fact that the commuter rail service studied also facilitates the movement of freight rail services. The resulting effect is captured in property value decreases of approximately 20 percent within 400 feet of the station.[323] It is impossible to differentiate between the impact attributable to commuter rail service proximity and that attributable to freight rail service. It can be assumed that the decline in value is due to noise and per or nuisance effects of freight services, yet these effects are not quantified.

San Francisco Gatzlaff and Smith (1993) observed that the total impact on property values is measurable in the San Francisco Bay area, but small, "...property prices and rents were raised in certain station areas [and that more recent development appears to occur more in] the urban CBD than suburban commuting stations".[324]

BART, a newer transit system has attracted development around its stations that is of a higher quality than development occurring away from stations.[325] It therefore becomes difficult to determine whether property value premiums are a result of station location or higher quality development.

Landis et. al. examined three Counties accessible by BART. Access provided a price appreciation for property in East Bay residential neighborhoods of $1.96-$2.26 per meter closer to the rail transit station, or approximately $70,000 depreciation in property value at 35kms from the station.[326] They also concluded that homes close to freeway interchanges experienced property value reductions of $2.80-$3.41 per meter.[327] Interestingly, the effect of proximity to freeway interchanges on property value is not only negative, but is an impact 30 percent greater than that of transit station proximity effects.

It can be concluded that property values are generally positively influenced by station location. However, a negative impact is generally explained by characteristics of the system itself or the relationship between the station and the overall transit orientation of the neighborhoods and business districts it serves.[328]

Los Angeles Fejarang studied the Los Angeles Metro Rail to determine the effects of transit station *announcement*[329] captured in the form of property values. The analysis examined 152 commercial parcels both before and after the announcement. Prior to an *announcement*, property values between expected station areas and expected non-station areas were insignificant. However, the period of realization illustrates a dramatic change. Areas both within and without the proximity of a metro station area realized property value gains of 78 percent and 38 percent respectively.[330] In hard currency terms, properties near rail have a mean sale price per square foot of $102 compared to properties away from rail with a mean sale price per square foot of $71, a difference of 30 percent.

New York City Alex Anas examined rail transit in the New York Metropolitan Area to quantify property value effects in regard to station locations. He studied a total of 18,649 parcels by building class and borough.[331] Anas stated that 1 per 3 of a property parcel's value could be lost if located one quarter mile away from a transit station measured by the shortest path walking distance.

Property values are also affected with respect to transit quality. In a recent study, Anas determined the change in property values resulting from an increase in the frequency of transit service from five minutes to two and a half minutes.[332] This increase in frequency provides a residential property value premium of $24 per year and a commercial property value of $0.06 per square foot per year.[333]

Anas examined other modes of transport in this study and found that "only subway improvements have a net positive effect on central area housing values", all other mode improvements move residents away from central areas to suburbs, thus decreasing central property values.

Observing the micro effects of station location, i.e. the immediate station area, negative attributes of the station and neighborhood generated lower property values and positive attributes created property premiums.

Stated Preference Study

The stated preference method has only recently been applied to neighborhood attribute valuation. A Canadian study of residential location in Calgary implemented this methodology and showed that individuals attached significant value to light rail station proximity.[334]

A stated preference survey was designed to measure the impact of various location specific neighborhood attributes. The survey instrument measured the impact of money cost per month, the number of bedrooms, distance to work, distance to shopping, and proximity to light rail transit station.

The survey presents hypothetical bundles of location attributes and asks respondents to choose preferred locations from sets of alternatives. By observing how willingness to pay in money cost per month changes with respect to proximity to transit, controlling for all other variables, an estimate of the value of transit proximity is estimated.

The results of the experiment suggest that survey respondents value walking distance to transit at C$217 per month (or about $150 in US dollars). Other results indicate that being within walking distance to light rail is worth 6.8 percent of respondent's income.

Conclusion

Up to this point, we have shown the historical basis of North American urban development paradigms from the early years of American settlements up to the new urbanism of recent years. The following questions emerge from this discussion:

> Do new visions (TOD and the new urbanism) for neighborhood development provide increased benefits compared to traditional auto-oriented neighborhoods?
>
> Can public policy contribute to the development of livable neighborhoods?
>
> Does transit access provide livability benefits to neighborhood residents?

While answers to these questions are beyond the scope of this review, the literature on livable communities suggests that:

> Access to transit services affects neighborhoods in ways that increase their value to residents.

Neighborhoods with transit access are generally preferred to auto-dependent neighborhoods.

The benefits of living in a transit accessible, livable neighborhood can be measured by transportation cost savings and property value increases.

New Research

In reviewing the available literature regarding livable neighborhoods and the role of transit, Hickling Lewis Brod Economics has drawn several conclusions that guided the planning and execution of this research. The literature suggests that transit plays a vital role in neighborhoods served by high quality rail transit. The impacts of transit include:

lower transportation expenses

higher property values

changes in development patterns

Transportation and land use are intertwined, each influencing the development of the other. Providing high quality transit together with development policies that allow or encourage transit oriented development, can influence land use patterns toward; higher densities, better pedestrian environments, and mixed use developments clustering around rail transit stations.

Current development patters have a historical basis in terms of social preferences (often reacting to urban problems), technological change, and government policy decisions. These influences have caused sprawling development patterns using large amounts of formerly open space, ineffective transit access, increasing pollution, and excessive congestion. These unfortunate conditions and changing government policies suggest that new urban development paradigms may be emerging to deal with these conditions. These modern urban problems may be addressed by building more livable neighborhoods which are transit oriented, higher density, and mixed use.

The concept of neighborhood livability encompasses two main types of potential benefits. The first benefit category is resource savings which encompass items included in typical benefit cost analysis such as reduced vehicle miles traveled, time savings, vehicle operating costs and the like. Although these types of benefits are valued in conventional benefit cost analysis, they may be underestimated by the failure to account for the effects of changes in urban form on resource consumption that occur apart from the actual use of the transit system (lower auto costs resulting from higher densities around stations). To the extent that consumers are aware of the resource savings, this value should be capitalized into property values around transit stations.

The second benefit category refers to value judgments about ideal urban form and how individuals should or desire to live. Sources for these value based benefits include public policy decision makers, academics and planning professionals, and individual consumers. To the extent that these benefits accrue to those affected by transit station location and access, these benefits should also be capitalized into property values.

Approach to Further Study

Hickling Lewis Brod Economics's review of literature relevant to livable neighborhoods leads directly to a work plan that takes account of the lessons of previous experience. The conclusions of the literature motivate the suggested approach. To assess the value of transit oriented communities, hedonic methods provide an accepted means of measuring residents willingness to pay for these attributes.

The hedonic study estimates the value of transit to station area residents. The estimate will include both transportation benefits and any non-use benefits of transit derived from neighborhood form and general livability. Currently, there is no sure way to separate these two effects. Therefore, counting the property value impact additively in a cost benefit analysis would double-count some of the benefits from transit. Disentangling these benefits is left to future research.

Introduction to Hedonic Methods Hedonic price estimation is performed using multiple regression techniques where the change in property values are a function of community amenities (distance to transit stations) and other social and economic variables. The regression coefficients are then used to calculate the implicit *marginal* prices (or shadow prices) of the amenities.

This research used the hedonic approach, both because the method is well established in the literature and because Hickling Lewis Brod Economics has developed, with Criterion Inc. a new method for quantifying model inputs. This approach makes use of Geographical Information Systems (GIS) techniques to measure, as accurately as possible, the walking distance to transit; the key variable in hedonic models of transit-oriented neighborhoods. Data for hedonic models can be purchased at relatively low cost and calculations for walking distance to transit can be readily made using GIS.

Benefits of Hedonic Models Hedonic methods are well established in the economics literature as a means of valuing amenities. The benefit of using hedonic price models is that they quantify amenity values by estimating shadow prices, that otherwise are not known. Clean air, for example, is an untraded commodity. However, its value or price can be observed by differential property values in areas with clean air compared with high pollutant areas, all other things being equal. It is evident that clean air is valuable, but at what price?

Hedonic price models can be employed to justify that the difference in property values, accounting for all other factors, *is* the price of clean air. Freeman (1993) stated that hedonic price models "have the capability of capturing the value of all of the possible effects of changes in environmental quality at a housing site in a single number".[335]

Limitation of Hedonic Methodology Although hedonic price models can quantify intangible values, they have some distinct limitations. They are limited by the assumptions of full information, perfect competition, and market clearing prices. However, each individual's transaction price represents the true stated value of the property. Hedonic models with sufficient data surmount these limitations and approximate the competitive market.

Structure of a Hedonic Study House values and property values are not mutually exclusive in the stated price, it is therefore necessary to control for structural characteristics that differ across the sample. A general hedonic equation examines the relationship of a dependent variable with all of its related independent variables. If property value is the dependent variable, it is necessary to account for all variables which influence property value. Property values are influenced by land size, house size, neighborhood accessibility, neighborhood amenities, and population, to name a few. The model is comprised of all structural characteristics of the dependent variable, characteristics of the neighborhood and per or environment, characteristics of the amenity being analyzed and an error term. The typical hedonic equation takes the form:

$$P_h = P_h(S_i, N_i, Q_i)$$

Where, S_i represents the structural characteristics of the property, N_i represents the characteristics of the neighborhood per environment, and Q_i represents the location specific amenities and *i* represents the individual property at the *ith* location.

The hedonic price function can take on various functional forms including a linear or exponential form. Functional form should be selected according to the goodness-of-fit criteria and on the basis of economic theory. Armstrong (1994) employed the following linear hedonic function to examine the value of transit accessibility in the Boston area:

$$P_h = \alpha + \sum_{i=1}^{I} \beta_i B_i + \sum_{j=1}^{J} \beta_j S_j + \sum_{k=1}^{K} \beta_k T_k$$

$$+ \sum_{l=1}^{L} \beta_l A_l + \sum_{m=1}^{M} \beta_m E_m + u_h$$

where

P_h = *h*th observation of housing prices

α = intercept term (constant)

β = estimated coefficients for each characteristic

B_i = structural attribute variables

S_j = site attribute variables

T_k = the local service provision and costs variables

A_l = locational and accessibility variables

E_m = local environmental impact variables

u_h = stochastic disturbance (error term)[336]

Estimates of the coefficients of each independent variable are calculated to explain the relationship between the dependent (property values) and the independent variables (characteristics). The sign of the coefficient defines the direction of change that an independent variable causes in the dependent variable. A positive coefficient means that as the independent variable rises, so does the dependent variable. The size of the coefficient defines the magnitude of the relationship.

Coefficients may be defined in terms of elasticities or marginal implicit prices. A linear specification returns coefficients that in terms of implicit marginal prices (shadow prices) while logarithmic specifications return coefficients in terms of elasticities. For example, if the coefficient of distance from a transit station location is 0.067, a 10 percent change in distance, all other things held constant, results in a 6.7 percent change in property value. Coefficients stated as marginal implicit prices explain absolute changes in the dependent variable, i.e. a coefficient of distance from a transit station is -15, a 1 unit increase in distance from transit results in a $15 reduction in property values on average.

Measuring Distance with a Geographic Information System (GIS)

This study used an approach from the literature of hedonic property value studies of transit station areas. Previous studies have focused, primarily, on large data sets spread across metropolitan areas. The San Francisco research by Landis et al (1995) used county wide data and measured property value impacts over 30 km from BART. The research in this study takes a more local point of view using a single station area within a one mile radius of BART.

This research seeks to find neighborhood level impacts of transit access. The hypothesis of this research is that transit improves the livability of transit oriented neighborhoods, produces benefits across the neighborhood, whether or not a particular resident uses transit. By finding a property value benefit with transit access regardless of use helps to confirm our notion of a neighborhood benefit apart from transit use. The property value premium represents a willingness to pay for transit proximity.

This approach uses data collected from real estate transactions and local government Geographical Information System (GIS) data. Previous hedonic studies have used geographical distance to measure property value effects of transit access. The purpose of a housing hedonic study is to measure the impact of some property attribute on property value, resulting in an estimate of the willingness-to-pay for that attribute. The property attribute that must be measured in a transit access study is the actual walking distance to the transit station holding all other attributes constant.

The typical solution to generating data on walking distance to transit is to use point to point, straight line distance from each property parcel to the transit station. This measurement is typically made by simply assuming that residents can walk in a straight line from their home to transit. This is never a truly accurate estimate of walking distance because streets do not always lead directly between two points. Some streets curve, meander, or dead end while other streets are cul-de-sacs which do not allow access at the shortest distance between points. Studies which use geographical distance to approximate walking distance to transit will miss some significant variations between properties. Some properties that are physically closer to a transit station than another property may be several minutes further away by actual walking distance, depending on the efficiency of the street grid.

The use of Geographic Information Systems (GIS) is a major innovation over the typical hedonic methodology applied to transit station areas, both in accuracy and in cost. GIS allows the precise measurement of the most important variable in the hedonic equation; namely, actual walking distance to the transit station. Hickling Lewis Brod Economics employed Criterion Inc., an urban planning firm, to generate actual walking distances from parcels to transit and to map the results on a property by property basis (see Figure 5.9). The GIS contains detailed information regarding the street grid in a given area and specifies each property parcel within the area in question. By calculating the shortest street distance from each parcel to the transit station, detailed data regarding the true variable of interest, walking distance to transit, is accurately specified. This improvement makes running hedonic property value studies involving access much easier.

Previous studies have attempted to use actual walking distance to transit in hedonic models. However, these studies have been time consuming and expensive since walking distance must be measured by "hand"; either by measuring the street distance on a map or by actually walking the route and recording the measured distance. GIS software makes this calculation a transparent exercise. All that is needed to replicate our methodology is a GIS database that includes geo-coded property parcels and a real estate database that includes property characteristics and matching geo-codes for the property parcels. Using standard GIS software and some specialized GIS manipulation products[337] distances are calculated for each property. Multiple regression techniques are applied to the databases producing estimates for the impact of transit proximity (measured by walking distance) on property values.

Another major benefit of GIS technology is the presentation of the results. GIS products produce visual mapping results which show the effects of transit access on property values directly on a map. Each parcel can be color coded to show the property value impact of its location relative to transit. This presentation is clear, concise and more visually interesting than a regression results table (see Figure 5.9).

San Francisco Bay Area Rapid Transit (BART) Study Area The study area radiates from the Pleasant Hill BART station along the yellow line. This station area is well outside of San Francisco proper, lying east of Berkeley in a low-moderate density suburb within Contra Costa county. The neighborhood is made up of mostly single family homes with some office, shopping, and multi-family residential development closer to the station. The area is made up of middle to high income residents at nearly $60,000 per household. Average home values in the station area are nearly $250,000.

This station is well established, having opened in 1973, suggesting that nearby residents should be well accustomed to the available transit options. Property values in the area should have fully adjusted to reflect the neighborhood development impacts and transportation benefits from BART service.

The area of transit impact is assumed to be approximately one mile from the Pleasant Hill BART station. This impact area is consistent with transit impact findings in previous studies.[338] While previous studies have found property value impacts outside a one mile radius (Landis, et. al.), this study seeks to focus on station area impacts. This impact area corresponds to walking (or biking) distance to transit. Most station area residents will not be willing to walk much more than a half mile to transit. For this reason, areas within one mile of BART should encompass the areas most likely to show changes in neighborhood structure to facilitate transit-oriented development.

Two primary sources provided data for the Pleasant Hill study area. Contra Costa county provided GIS mapping data which allowed the calculation of walking distances from each parcel to the Pleasant Hill station. Home sale price data was purchased and matched to parcel numbers in the GIS database.

New York City Queens Study Area Decidedly more urban than the other study areas, the Queens study area focused on three New York City MTA Subway Stations. These were Forest Hills, 67 Ave, and Rego Park; all within the neighborhoods of the same names, Forest Hills and Rego Park.

These stations fall along the E, F, R lines which travel to uptown Manhattan before spitting off to downtown, Harlem, and the Bronx. The G line travels to downtown Brooklyn. Forest Hills is served by all four lines while the other station are served by the G and R lines.

While the New York City Subway system is much older and in worse condition than the other systems in this study, the scope and mobility offered by the system is unmatched in the United States. New York City neighborhoods warrant detailed study by virtue of their transit dependence. If mobility on the transit system provides benefits to residents, Forest Hills should display stronger property value impacts by virtue of its superior access to major subway lines.

Forest Hills is the highest priced neighborhood in the study with average home values around \$390,000. The homes nearest 67 Ave are less costly at about \$226,000, and Rego Park lowest at just under \$200,000.

Household income is also highest in Forest Hills at nearly \$60,000 followed by 67 Ave at about \$50,000 and Rego Park at about \$44,000 per household. Median household income may appear low given housing prices. The likely explanation is that lower income individuals live in the numerous apartment buildings in the station areas.

Data for the New York study were provided by the City Planning Office and TRW. The planning office provides GIS mapping data on CD-ROM for every Borough in New York City. The real estate database from TRW provided sale prices for homes in our study area.

Unfortunately, the real estate data for New York was significantly less detailed than for other areas. Data regarding home size and other physical characteristics were unavailable. This limitation was mitigated by using data aggregated by census district. This approach provides average home characteristics at the level of detail of a few blocks.

Portland, Oregon MAX Study Area The analysis of Portland's MAX light rail station areas tested three stations along the East Burnside corridor: the 148th Avenue, 162nd Avenue, and 172nd Avenue stations. These three stations are less than a mile apart, creating a heavily transit served neighborhood. The light rail system in Portland, primarily uses existing right-of-ways down major arterial streets. Land-use surrounding these stations is dominated by single family detached, moderately priced homes with relatively small amounts of multi-family residential and civic (schools and parks) buildings. The average home value in the station areas within one mile of the three stations is about \$95,000.

The GIS database for the Portland area was collected from the metropolitan planning agency (Metro). This data was linked to a database of property tax assessment records.

BART – Pleasant Hill Study The BART study results are strongly significant and show that BART station proximity is a key determinant of property values in Pleasant Hill. The research shows that single family homeowners are willing to pay, on average, nearly \$16 in home price for each foot *closer* to BART within the study area (see Figure 5.8). These values reflect the neighborhood and transportation benefits derived from

BART access. Alternatively, homeowners are willing to pay nearly $8 in home price for every foot *further* from the freeway interchange nearest the study area, most likely reflecting the noise, pollution, and unsightliness of development near freeways.

The value of an average single family home in the Pleasant Hill Station Area is $22,767 greater due to its proximity to BART. For the 939 single family homes within a 1 mile radius of this station, the net property value impact is $21.4 million. Alternatively, neighborhood property values are about 10 percent greater due to the existence of the BART station in Figure 5.8 shows how property values decay with distance from BART. A graph is given for 2 bedroom homes and 3 bedroom homes and shows good fit between predicted and actual values.

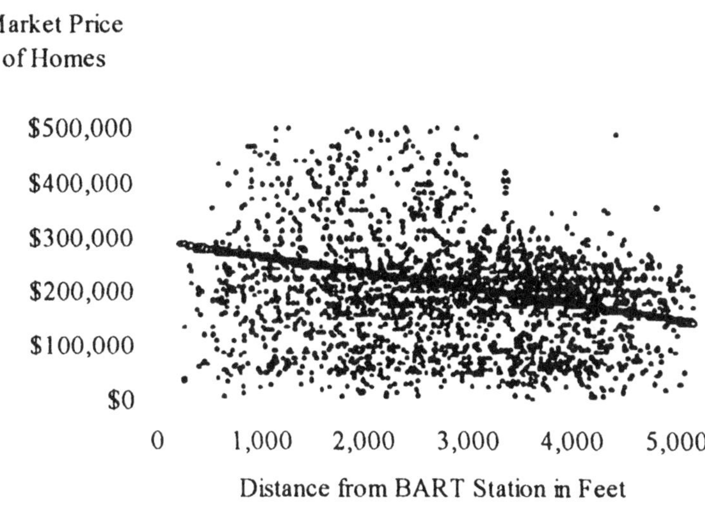

Figure 5.8 Property Values and Transit Stations

Source: Hickling Lewis Brod Economics, Inc., 1996.

Assuming that this station area is representative of those in the San Francisco area,[339] the property value impact from BART for single family homes is over $725 million. This estimate should be regarded as a lower bound given the relatively low-density development around the Pleasant Hill study area.

Model Specification A hedonic model is used to isolate the effects of proximity to BART on property values near the Pleasant Hill station. The model is specified to include a mix of home characteristics and transportation characteristics to account for as much property price variation as possible.

Data on home characteristics includes such items as number of bedrooms, number of bathrooms, size of the home in square feet, lot size, age of the home and other items. All of these variables are not included in the model because many of these variables are highly correlated with the others. For example, the number of bedrooms will be correlated with size of the home since bigger houses have more bedrooms.

Testing various regression equations found that home size and age of the home accounted for most of the variation in home values due to home characteristics. The variables included in the final regression are:

home age in years (Home Age)

home size in square feet (Home Size)

walking distance to BART station (Distance to BART)

distance to highway interchange (Distance to Highway)

The best regression, which accounts for over 80 percent of the variation in property values in our sample, has the following specification:

$$HomeVal = \alpha + \beta_1 Dist\_to\_Bart + \beta_2 Dist\_to\_Hwy$$
$$+ \beta_3 HomeAge + \beta_4 HomeSize + error$$

The primary coefficient of interest is β_1 which is the change in home value from a one foot change in walking distance to the BART station. A positive coefficient means that transit has a negative impact on property value while a negative number means transit proximity enhances property values. The hypothesis of this study, which is confirmed in the results (see Table 5.5) is that this coefficient will be negative and significant showing that transit access provides economic value that is capitalized in local property values.

Analysis of BART Results The regression results in Table 5.5 show that for homes in the study area, BART access is worth $15.78 more for every foot closer to the station on average. This means that an average home in our study area would be worth over $15,000 more if it were 1000 feet closer to BART than its original location. Interestingly, closeness to highways has a negative effect on housing values within our study area. The regression shows that homes further from a highway interchange are worth $7.94 more on average for every foot further from the freeway interchange.

The home characteristics variables are extremely good indicators of home values. Building size is the most important determinant of home prices with a value of about $100 per foot. Home age tends to reduce property values by about $443 per year. All explanatory variables in this regression are highly significant.

Table 5.5 Regression Results with Linear Specification

Dependent Variable : Home Sale Price in 1995 Dollars

Variable	Coefficient (t-statistic)
C	143,504.9 (8.70)
Home Characteristics	
Age of Home	-422.79 (-2.48)
Size of Home	100.39 (21.14)
Transportation Characteristics	
Distance to BART	-15.78 (-5.79)
Distance to Highway	7.94 (3.15)
All coefficients are significant at the one percent level	
Summary Statistics	
Number of Observations	263
R^2	.81
Mean Dependent Variable	249,848.4
F--Statistic	272.999

The logarithmic specification in Table 5.6 replicates the previous linear specification using natural log transformations of the variables. The results show the same relationships expressed in the previous regression, but expresses them in terms of elasticities.

The interpretation of the coefficients is that the value equals the percent change in home sale price given a 1 percent change in the independent variable. For example, a 1 percent increase in distance from BART results in a 0.22 percent reduction in home price. The interpretation of the other coefficients follows similarly. A one percent increase in distance from the highway leads to a 0.10 percent increase in home sale price. A one percent increase in home size leads to a .62 percent increase in sale price while a 1 percent increase in home age leads to a 0.05 percent decrease in home sale price.

Table 5.6 Regression Results with Logarithmic Specification

Dependent Variable : Log (Home Sale Price) in 1995 Dollars

Variable	Coefficient (t-statistic)
C	9.04 (19.72)
Home Characteristics	
Log (Age of Home)	-0.05 (-3.34)
Log (Size of Home)	0.62 (18.19)
Transportation Characteristics	
Log (Distance to BART)	-0.22 (-5.63)
Log (Distance to Highway)	0.10 (3.61)
All coefficients are significant at the one percent level.	
Summary Statistics	
Number of Observations	263
R2	.77
Mean Dependent Variable	12.38
F--Statistic	216.05

Figure 5.9 shows the property value impacts on a Geographical Information System (GIS) map. The GIS maps the locations of parcels containing single family homes and calculates the distance to the transit station via available streets to estimate actual walking distance to BART. The parcels have been color coded to show how forecasted home prices decline as their distance from BART increases.

Evidence of Non-User Benefits The results of our Pleasant Hill area research confirm a large and significantly positive impact of access to BART on property values around the station. These property value impacts reflect an array of benefits from transit access that this study cannot fully delineate. Some of the premium paid for proximity to transit compensates for the reduced travel costs. This compensation is measured by the benefit from trips actually taken.

However, there may be a non-use benefit which is evidenced in two ways. First, by the fact that many people who live near transit are willing to pay a property value premium, yet do not use transit. And second, that the amount of the observed property value premiums are too large to be explained by user benefits. These are more fully illustrated below.

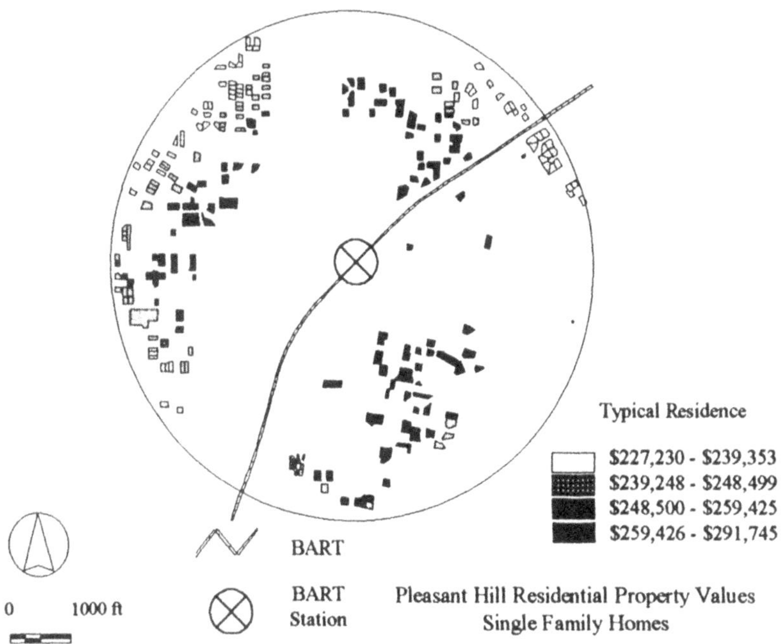

Figure 5.9 Geographic Information System Map of Study Area

Source: Hickling Lewis Brod Economics, Inc., 1996.

First, consumers pay a premium regardless of transit use. There are many individuals in the study area who pay premiums in housing prices in excess of $20,000 to live near transit but will never use transit. This willingness to pay the premium must reflect some value of transit proximity that accrues to residents regardless of transit use. Second, the value premium is too large to represent user benefits. To illustrate this point, consider the following scenario.

Two residents of the Pleasant Hill Neighborhood are regular BART users who walk to the Pleasant Hill station. One resident lives ¾ mile from BART while the other lives ½ mile away.

The logarithmic regression results in Table 5.6 show that moving ¼ mile closer (3 per 4 mile to ½ mile) to the Pleasant Hill BART station results in $18,000[340] in added property values, holding all other property characteristics constant. The $18,000 in property value leads to about $130 per month in additional mortgage costs at 8 percent interest for 30 years. This is the observed monthly willingness-to-pay to live ¼ mile closer to transit.

The walking time for a ¼ mile trip is about 5 minutes.[341] At that rate, the resident ¼ mile closer to transit, saves about 10 minutes per day (two trips) or about 3.3 hours per month (20 travel days). Even at an upper bound estimate for value of time of around $20 per hour for time savings,[342] a resident would be willing to pay only $66 per month for the time savings of living ¼ mile closer to transit. This is only 50 percent of the observed willingness to pay.

In fact, the value of time would need to be about $40 per hour; higher than nearly every estimate found in the literature for intra-urban commuting trips.[343] Therefore, the observed willingness to pay for transit station proximity most likely includes non-user benefits of transit. These non-user benefits likely amount to at minimum, 50 percent of the observed property value premium.

New York City MTA–Queens Study The study of New York City subway station areas focused on three neighborhoods in the Borough of Queens. These neighborhoods are transit oriented and enjoy easy transit access to uptown and downtown Manhattan and Brooklyn. The transit dependence New York Neighborhoods and the high degree of transit mobility on the New York Subway system suggest that these stations should provide large benefits to residents within walking distance.

The results for these station areas show high levels of benefits for residents within walking distance. The aggregate data set shows that, on average, home prices decline about $23 for every foot further from the subway stations (see Table 5.7). This value represents the average willingness to pay for proximity to these subway stations.

The value of an average home within these subway station areas is about $37,000 greater than a home outside the station areas. For the 2700 single family residences in the station areas, the net property value impact of proximity to the subway is approximately $100 million or about $30 million per station. This estimate refers to benefits to single family home owners. A majority of residents (80–90 percent) actually live in multi-family housing units.

Table 5.7 Regression Results for Queens Stations, Aggregate Data

Dependent Variable Home Sale Price in 1996 Dollars

Variable	Coefficient (t-statistic)
C	103,747 ((5.81)
Home and Demographic Characteristics	
Lot Size	48.08 (21.03)
Forest Hills Indicator	28,992.27 (2.607)
Median Income	1.89 (5.299)
Transportation Characteristics	
Distance to Station	-23.49 (-7.023)
Distance to Highway	5.93 (3.034)
All coefficients are significant at the one percent level	
Summary Statistics	
Number of Observations	1738
R^2	.424
Mean Dependent Variable	293,076.1
F–Statistic	254.73

Analysis of the New York Results These results confirm and mirror the results from Pleasant Hill near San Francisco. The relatively high level of transit service (4 lines vs. 2 lines) at the Forest Hills station suggests that the property value impact should be stronger at this station.

Simply showing a larger coefficient on the distance variable in a Forest Hills regression is insufficient to show stronger property value impacts. The higher prices of the Forest Hills properties relative to the other station areas means that the larger coefficients of the distance variable would be expected, even if the *relative* property value impact were the same in all station areas. A log-linear regression, because it estimates elasticities, is estimated to test whether better service increases the strength of the property value impact.

Log linear regressions were constructed for the Forest Hills station area and for the combined 67 Ave and Rego Park station areas. The results confirm that the Forest Hills station area provides much stronger property value impacts than the other areas. The results of the log-linear model show that for every 1 percent increase in distance to the station, property values in Forest Hills decline 0.3 percent. The other station areas show declines of only about 0.1 percent as distance from transit increases by 1 percent.

The strength of the property value benefits near Forest Hills may be the result of the higher transit service relative to the other stations. An alternative explanation is that the immediate station area in Forest Hills (possibly as a result of demographic factors) may be more pleasant and, consequently, a more desirable location compared to the other areas.

MAX–Portland East End Study The initial application of this methodology was conducted for Portland, Oregon's light-rail transit line. This area was chosen because Portland is widely known as a city that supports the concept of building livable neighborhoods around transit. Zoning and transportation investments are specifically geared toward orienting neighborhoods to transit rather than the automobile. The region is a focal point in the public policy debate regarding transit-oriented development, having one of the most elaborate plans for implementing transit-oriented policies in the nation. The Portland region has not fully implemented the transit-oriented development concept, but has implemented several policies, urban growth boundaries and complementary zoning for mixed use and infill projects, that serve to spur more transit friendly neighborhoods.

Portland results are also less clear because the transit system is light rail (streetcar type transit) which rides on existing street right-of-ways. This is a drawback in terms of transit system performance. For transit oriented neighborhoods to develop, transit must be an attractive alternative to automobiles. Light-rail transit may not fit this description because it is typically slower than auto travel and does not have the capacity of heavy rail (subway systems such as in New York, Washington DC, San Francisco among others). The lower performance characteristics of light rail transit may limit their ability to provide significant value observable in property values.

Preliminary results from hedonic analysis of Portland, Oregon light rail transit station areas were problematic with the initial sample providing results contrary to our expectations. The results are presented in Table 5.8.

Analysis of Portland Results A number of factors specific to the Portland area and data set help to explain the anomalistic results and, in fact, suggest some interesting implications for getting the highest value out of transit. One explanation is that light rail vehicles are slow and have less capacity compared to heavy rail transit. The service characteristics of light rail are far below the service characteristics of most heavy rail systems leading to the expectation that the property value impacts will be much weaker than for a heavy rail transit system.

Table 5.8 Results for Portland, Oregon Transit Station Areas

Dependent Variable : Assessed Property Values, 1994

Variable	Coefficient (t-statistic)
C	41431.83 (26.54)
Home Characteristics	
Age of Home	-506.47 (-27.44)
Size of Home	39.74 (77.7)
Lot Size	4.59 (31.0)
Residential Zoning (1 = Yes)	2777.84 (2.41)
Transportation Characteristics	
Distance to Light Rail	1.41 (7.48)
All coefficients are significant at the one percent level	
Summary Statistics	
Number of Observations	4,170
R^2	.69
Mean Dependent Variable	93,211.54
F–Statistic	1548.65

A much more interesting and testable explanation is that the impacts of transit proximity and highway proximity are conflicting in the Portland data. As the BART study showed, proximity to a highway is strongly negative for property values. Portland's light rail line runs down a major arterial street implying that the negative effects of proximity to highways are conflicting with the positive impacts of the light rail transit line.

This hypothesis was tested by looking for positive transit access impacts further from the light rail stations. Regressions were run restricting the data set to properties successively further from the transit stations and the major roadway. The results in Table 5.9 suggest that transit access increases property values as long as properties are over 2000 feet from the major roadway and transit line. The sign on the distance to transit variable becomes negative when properties greater than 2000 feet from both transit

Table 5.9 Results for Portland Station Area for Distances > 2500 ft.

Dependent Variable : Assessed Property Values, 1994

Variable	Coefficient (t-statistic)
C	49924.61 (18.02)
Home Characteristics	
Age of Home	-477.47 (-20.16)
Size of Home	40.04 (61.55)
Lot Size	4.35 (21.49)
Residential Zoning (1 = Yes)	2567.98 (1.27)
Transportation Characteristics	
Distance to Light Rail	-.757 (-2.00)
All coefficients are significant at the five percent level	
Summary Statistics	
Number of Observations	2,660
R^2	.69
Mean Dependent Variable	94,792.71
F--Statistic	987.04

and the major roadway are included in the sample. In fact, the coefficient on distance to transit becomes significant only when properties past 2500 feet are included. The following table presents results for Portland for properties over 2500 feet from the light rail station and major roadway.

The results of the distance restricted regression show that property values decline as distance to light rail increases within the included sample. However, the coefficient suggest a much smaller property value impact in Portland than for BART. This is not surprising given the lower performance of light rail in Portland and the much lower property values generally in the Portland region compared to San Francisco.

The coefficient on the distance variable suggests that property values increase by about $0.76 for every foot closer to light rail within the 2500 feet to 5280 feet distance to transit range included in the sample. Controlling for all other variables, homes 1000 feet closer to transit are worth about $760 more than other homes, on average. While statistically significant, this property value premium is small compared to the results from San Francisco.

Policy Implications

The immediate result of the Portland regression is that not all transit stations provide proportionately the same benefits. The results from BART suggest strong property value benefits from transit, while Portland only shows benefits to properties more than ½ mile from transit. The BART station at Pleasant Hill is located near where the transit line breaks away from the freeway right of way providing distinct data for distance to freeway and distance to BART. Portland light rail, running down a major arterial road with relatively low performance, provides no such opportunity.

These results suggest that building transit lines on freeway or major road right-of-ways sacrifices the neighborhood, livability benefits of transit. Transit systems built along freeways will most likely produce the transportation user benefits normally associated with transportation investments. However, the results of this study suggest high quality heavy rail transit, integrated into the structure of a neighborhood and outside the negative impact areas of major freeways, can provide benefits in excess of the transportation user benefits.

Notes

243 Transit-Oriented Development refers to areas served by multiple transportation modes (auto, transit, and pedestrian) and characterized by higher residential density and mixed land uses. See Calthorpe, Peter, *The Next American Metropolis,* (New Yoirk: Princeton Architectural Press, 1994).

244 Owens, Peter, "Defining a Livable Community", unpublished draft, (Fall 1994).

245 Calthorpe, Peter. op. cit., (1993).

246 Lowe, Marcia, "Alternatives to the Automobile: Transport for Livable Cities," *Paper 98*, (Washington, D.C.: World Watch Institute, 1990).

247 Rybczynski, Witold, *City Life: Urban Expectation in a New World*, (New York: Scribner, 1995), p. 64.

248 Ibid., p. 67.

249 Ibid. pp. 94-100.

250 Ibid. pp. 100-101.

251 Fishman, Robert, *Bourgeois Utopias: The Rise and Fall of Suburbia*, (New York: Basic Books, 1987).

252 Boyer, Christine, *Dreaming the Rational City: The Myth of American City Planning*, (Cambridge: MIT Press, 1983).

253 Rybczynski, pp. 111.

254 Hall, Peter, *Cities of Tomorrow: An Intellectual History of Urban Planning and Design in the 20th Century*, (London: Basil Blackwell, 1988).

255 Weyrich, Paul and William Lind, "Conservatives and Mass Transit: Is it Time for a New Look?", (Washington DC: Free Congress Foundation and APTA, preprint, 1996).

256 Kunstler, James Howard, *The Geography of Nowhere: The Rise and Decline of America's Man-Made Landscape*, (New York: Simon and Schuster, 1993).

257 Rybczynski, op. cit., p. 159.

258 Jeanneret, Charles-Edouard (aka Le Corbusier), *When Cathedrals Were White*, (New York: McGraw-Hill, 1964).

259 Bellah, Robert N. (with R. Madsen, W. Sullivan, A. Swidler, S. Tipton), *The Good Society*, (New York: Vintage Books, 1992).

260 Kunstler (1993), p. 105.

261 Downs (1994), pp. 27-8.

262 "World War II and the American Dream", National Building Museum, 1995, exhibit documentation.

263 Kunstler (1993), p. 114.

264 Typically by imposing prices equal to the marginal imposed costs.

265 Laffont, Jean-Jacques, *Fundamentals of Public Economics*, (Cambridge: MIT Press, 1989), pp. 10-12.

266 Kitamura, Ryuichi, "Life-Style and Travel Demand", *A Look Ahead Year 2020. Transportation Research Board Special Report 220*, (Washington, DC: National Research Council, 1988), pp. 149-184.

267 Knight, Richard V., "Changes in the Economic Base of Urban Areas: Implications for Urban Transportation", *Future Directions of Urban Public Transportation*, (Washington, D.C.: Transportation Research Board Special Report 1983), pp. 54-58.

268 Real Estate Research Corporation, *The Costs of Sprawl,* vol. 1, (Washington, D.C.: United States Government Printing Office, 1974).

269 Audirac, Ivonne, Anne H. Shermyen and Marc T. Smith, "Ideal Urban Form and Visions of the Good Life, Florida's Growth Management Dilemma", *American Planning Association (APA) Journal*, (Autumn, 1990), pp. 470-482.

270 Weyrich, et al, p. 10.

271 Kunstler (1993), p. 79.

272 Jacobs, Jane, *The Economy of Cities*, (New York: Random House, 1960).

273 Rybczynski, p. 162.

274 Kain, John T., "Housing Segregation, Negro Employment, and Metropolitan Decentralization", *Quarterly Journal of Economics,* vol. 82, (1968), pp. 175-97.

275 Rybczynski, p. 164.

276 "Bye, Bye, Suburban Dream", *Newsweek*, May 15, 1995, p. 43.

277 Jacobs, Jane, *The Death and Life of Great American Cities*, (New York: Vintage Books, 1961).

278 Whyte, William H., City: *Rediscovering the Center*, (New York: Anchor Books, 1988).

279 Mumford, Lewis, *The City in History: Its Origins, its Transformations and its Prospects,* (New York: Harcourt, Brace and World. 1961).

280 Canadian Ministry of State for Urban Affairs, *Creating a Livable Inner-City Community: Vancouver Experience*, (Ottawa: Publisher Agency Press, 1976).

281 Oregon Department of Transportation, *Oregon Transportation Plan,* (1992), p. 5.

282 Ibid.

283 Mantell, Michael, Stephen T. Harper and Luther Propst. *Creating Successful Communities; A Guidebook to Growth Management Strategies,* The Conservation Foundation, (Washington, D.C.: Island Press. 1990).

284 Attoe, Wayne, ed., *Transit, Land Use and Urban Form*, Center for the Study of American Architecture, (Austin: University of Texas, 1988).

285 "Bye, Bye, Suburban Dream", *Newsweek*, May 15, 1995, p. 43.

286 Audirac, et al, p. 470.

287 Ibid., p. 474.

288 Rappoport, A., "Toward a Definition of Density", *Environment and Behavior*, (1975), pp. 133-58.

289 Moore, Terry and Paul Thorsnes, *The Transportation per Land Use Connection: A Framework for Practical Policy*, (Washington, DC: The American Planning Association, 1994).

290 Gordon, Stephen P. and Peers, John B., "Designing a Community for Transportation Demand Management: The Laguna West Pedestrian Pocket", in *Transportation Research Record,* No. 1321, Rideshare Programs: Evaluation of Effectiveness, Trip Reduction Programs, Demand Management, and Consumer Attitudes, (1991).

291 Bartik, Timothy J., "Measuring the Benefits of Amenity Improvements in Hedonic Price Models", *Land Economics*, vol. 64, no. 2, (May 1988).

292 Downs, Anthony, *Stuck in Traffic: Coping with Peak-Hour Traffic Congestion*, (Washington DC: The Brookings Institution, 1992), pp. 27-8.

293 Cervero, Robert, "Light Rail Transit and Urban Development", *Journal of the American Planning Association*, (Spring 1984), p. 134.

294 United States Bureau of Labor Statistics, *Consumer Expenditure Survey*, (1992).

295 Downs (1994). p. 37.

296 Calthorpe (1993), p. 27.

297 Holtzclaw, John. "Explaining Urban Density and Transit Impacts on Auto Use", Natural Resources Defense Council, January 15, 1991, in California Energy Commission Docket No. 89-CR-90.

298 Holtzclaw, John. "Using Residential Patterns and Transit to Decrease Auto Dependence and Costs", Natural Resources Defense Council, (June 1994).

299 Moffet, John and Peter Miller, "The Price of Mobility", Natural Resources Defense Council, (San Francisco, 1993).

300 TAI = number of 50 seat transit vehicles accessible to residents, in Holtzclaw, 1994.

301 Ibid., p. 21.

302 Green, R.D. and O.M. James, *Rail Transit Station Area Development: Small Area Modeling in Washington, DC*, Armonk, New York: M.E. Sharpe, 1993.

303 Mumford, Lewis, *The Culture of Cities*, (New York: Harcourt Brace, 1938).

304 For a recent example of a quantitative study regarding transportation policy in Portland, Oregon, see Annex 1: The LUTRAQ Model.

305 Pierce, p. 3.

306 Ibid.

307 Federal Transit Administration, *Livable Communities Initiative: Program Description*, (Washington, D.C.: United States Department of Transportation, September 1994), p. 3.

308 Hensher, David A., Peter Barnard, and Truong P. Truong. "The Role of Stated Preference Methods in Studies of Travel Choice", *Journal of Transport Economics and Policy*, (January 1988), pp. 45-58.

309 Lerman, Steve R., David Damm, Eva Lerner-Lamm, and Jeffrey Young, *The Effect of the Washington Metro on Urban Property Values*, Prepared for Urban Mass Transportation Administration, (July 1978).

310 Parsons Brinckerhoff Quade & Douglas, Inc. "Transit and Urban Form: A Synthesis of Knowledge". Prepared for Transit Cooperative Research Program -Transportation Research Board National Research Council, (October 1995).

311 Green, R.D. and O.M. James, *Rail Transit Station Area Development: Small Area Modeling in Washington, D.C.*, Armonk, (New York: M.E. Sharpe, 1993).

312 KPMG Peat Marwick, *Fiscal Impact of Metrorail on the Commonwealth of Virginia*, (November 1994).

313 Gatzlaff, Dean H. and Mark Smith. "The Impact of the Miami Metrorail on the Value of Residences Near Station Locations", *Land Economics*, vol.69 no.1, (February 1993), pp. 54-66.

314 Rice Center, Joint Center for Urban Mobility Research, 1987, "Assessment of Changes in Property Values in Transit Areas". Prepared for the Urban Mass Transit Administration.

315 Voith, Richard, "Transportation, Sorting, and House Values", *AREUEA Journal*, vol. 19 no. 2, (1991), pp. 117-137.

316 Voith, p. 123.

317 Voith, p. 136.

318 Cervero, Robert. "Light Rail Transit and Urban Development", *APA Journal*, (Spring 1984), pp. 133-147.

319 Ibid.

320 Armstrong, R.J., Jr. "Impacts of Commuter Rail Service as Reflected in Single-Family Residential Property Values". Paper presented at the

73rd Annual Meeting of the Transportation Research Board, (Washington D.C., 1994).

321 See Alex Anas (1995).

322 Armstrong, R.J., Jr., op. cit.

323 Ibid., pp. 96.

324 Gatzlaff, et al p. 55.

325 Landis, John et al, "Rail Transit Investments, Real Estate Values, and Land Use Change: A Comparative Analysis of Five California Rail Transit Systems". Prepared for Institute of Urban and Regional Development, University of California at Berkeley (July 1995).

326 Ibid., pp. 32.

327 Ibid., pp. 32.

328 Parsons Brinckerhoff Quade & Douglas, Inc. "Transit and Urban Form: A Synthesis of Knowledge", Prepared for Transit Cooperative Research Program-Transportation Research Board National Research Council, (October 1995).

329 'characterized by a series of federal, state, and local funding propositions that began in 1983 and was legislated in July 1988 for the purpose of transit investment'.

330 Fejarang, R.A., *Impact on Property Values: A Study of the Los Angeles Metro Rail,* Prepared for Transportation Research Board 73rd Annual Meeting, (January 1994).

331 Anas, Alex, *Transit Access and Land Value--Modeling the Relationship in the New York Metropolitan Area,* United States Department of Transportation, Federal Transit Administration, (September 1993).

332 Anas, Alex, "Capitalization of Urban Travel Improvements into Residential and Commercial Real Estate: Simulations with a Unified Model of Housing, Travel Mode and Shopping Choices", *Journal of Regional Science*, vol. 35 no. 3., (1995).

333 Ibid., pp. 371.

334 Hunt, J.D., J.D.P McMillan, and J.E. Abraham, "A Stated Preference Investigation of Influences on the Attractiveness of Residential

Locations", paper presented at the Transportation Research Board Conference, (1994).

335 Ibid., p.416.

336 Armstrong, Robert J. Jr., "Impacts of Commuter Rail Service as Reflected in Single-Family Residential Property Values", *Transportation Research Record* 1466, (1994), pp. 88-98.

337 Our study employed the services of Criterion Inc., a consulting firm specializing in urban planning and GIS applications. Criterion's proprietary software product, INDEX, allows for detailed analysis of GIS and geo-coded property value databases.

338 Pushkarev, Boris and Jeffrey Zupan, *Urban Rail in America: An Exploration of Criteria for Fixed-Guideway Transit*, (Washington, D.C.: United States Department of Transportation, 1980).

339 This is a conservative assumption considering that Pleasant Hill is an outer suburban station with lower residential densities than more urban stations. Other, more developed, station areas have higher total property value benefits because these areas have more homes that benefit.

340 Proximity coefficient implies moving 33 percent closer results in a 7.26 percent value premium (-0.33 * -0.22).

341 The walking time for 1 per 4 mile distance is found by using 3 miles per hour as the average walking rate. US Department of Transportation, *Characteristics of Urban Transportation Demand: An Update*, (Washington, D.C., 1988).

342 This value corresponds to the upper bound estimate found in: National Cooperative Highway Research Program, *Research Strategies for Improving Highway User Cost Estimating Methodologies*, (Washington, D.C., 1992).

343 An excellent survey of the value of time literature is found in: Button, Kenneth J., *Transport Economics*. (Cambridge: University Press, 1993).

6 Public Choice Analysis for Transit Policy and Planning

New Policy Directions

The foregoing suggests new directions for policy and planning, particularly in public transit. Policy makers and planners can make good use of Cost-Benefit Analysis if the analysis framework is informed by a sound understanding of actual public choices in the context of democratically determined budgets over a long period of time. Just as efficiency in economics is understood through the on-going analysis of actual buy-sell choices of the firm at the margin, so too is efficacy in public policy and project planning understood through the analysis of incremental budget decisions affecting transit services and marginal costs and benefits of these services to taxpayers.

Public Choice, Measurement, and Policy

The findings reported here challenge long-standing criticism of transit's efficacy. Persistent budgetary support for transit services in a very wide array of locations is undisputed and, it turns out, perfectly rational. We have identified an intuitively appealing and theoretically sound explanation of transit's success in local budgetary processes. Namely, affordable transportation is valued in every urbanized area in the United States. Private vehicle operation is the norm in the United States. But every community contains children, elderly people, and others who cannot safely drive and many who cannot afford cars. Local budgets extend transit services for these needs. Additionally, in a significant number of severely congested urban commuting corridors, rapid transit measurably improves the work trip for passengers and motorists alike. Also, transit fosters walkable residential, commercial, and campus concentrations that reduce the transportation costs of households, businesses, and institutions.

Against yearly budgetary outlays of $20 billion or so, the value of these outcomes to passengers and taxpayers is estimated to range from $45 billion to $60 billion per year. This range is based on several alternative measurements. The passenger benefits are calculated from revealed preference data such as price and income elasticities, product substitution effects, known values of time, and generalized (e.g., parking) costs. While more measurement is highly desirable and much needed, our estimates are probably conservative. The unit value of transit trips for people with disabilities are much higher, for example, but they are not reported here.

Transit's indirect benefits to local constituencies are estimated from known travel time savings to motorists in rapid transit corridors, from transportation costs in social service budgets, and from vehicle ownership data in neighborhoods with intensive transit. The constituency benefits are conservative. No commercial benefits of transit proximity are reported. Not reported are the economies of agglomeration associated with transit intensive cities.

Similarly, for diffuse benefits to the general public, only one externality is examined—harmful tailpipe emissions saved by lower VMT's per capita. Substantial other external auto costs that are avoided by transit use, such as avoided highway construction, remain uncounted in this book. Policy makers should be comfortable with the five to one ratio of return for transit subsidies. Theoretically, the economy reaps even larger benefits from transit.

If local transit professionals replicate these methods in their own communities, they can be confidant of finding useful information about the value of their transit services to passengers and other taxpayers. The measurements make sense and they are well grounded in economic theory. Customers that use transit to substitute for auto ownership are well aware of the dollars they save each year. The same goes for auto owners who use rapid transit to circumvent congested highways to go to work. Intuitively, motorists are conscious of the difference transit makes in congested travel corridors. Taxpayers who rarely use transit themselves willingly support transit for their children, their parents, neighbors, and others. With the measurements suggested in this book, local analysts can gauge the financial value of these indirect benefits. They can do so in a way that makes sense to local constituencies.

Moreover, these are measurements commonly used by practicing economists, market researchers, and small businesses to estimate returns for business enterprises. Product and service substitution is the essence of business growth. Businesses count on the public's desire to replace old shoes with new shoes, slower cars with faster cars, smaller houses with bigger houses. Business competition is little more than substitution.[344]

Chapter 2 of this book applies this principle of product substitution to determine the value of transit services to households, constituencies, and the general public. Like businesses, the test of transit efficiency or success is the cost and the value for the last dollar spent or the last unit of service purchased. The application of public choice principles to transit in this way approximates the principle of marginal cost pricing to reach efficient decisions.

Chapters 3, 4 and 5 apply conventional economic analysis tools to determine the value of transit. For the value to passengers, econometric analysis was used to determine the consumer surplus received by transit dependent households with incomes below the earned income tax credit ceiling—as an indicator of economic distressed households. Econometrics were also used to determine the efficient transit subsidy needed to offset implicit subsidies to use of congested highways. Substitution savings are calculated for transportation-related costs of public agencies and also for air pollution costs. Once again, these estimates are grounded in costs and benefits for the last transit trip.

Public Choice and Budgetary Incrementalism

At the end of the day, just as efficiency can be approached through marginal cost pricing in a multitude of businesses, efficiency can be approached through disjointed incrementalism in thousands of public sector budgets. This book documents a remarkably efficient public transit sector, revealing as it does a high benefit return for transit budgets. Local leaders, who by necessity practice incrementalism, can increase the efficiency of their decisions by explicit measurement of transit performance in its familiar public policy functions.

Most transit budgets are devoted entirely to affordable mobility services used by a minority of the local population in relatively small urbanized areas. The purpose of the service is maximum coverage for the population. Farebox returns are very modest compared to costs; usually less than 20 percent of costs. Our findings suggest, however, that the value of these services are typically much higher than their total costs. An application of the valuation methods reported in this book could help budget planners to better compare the value of expanded transit service to the value of other local priorities. Were local authorities to follow this suggestion, economists would predict expanded transit services to a point of decreasing marginal returns. Such studies would show that the benefits of affordable mobility call for the deployment of substantially more transit services than currently exist in urban areas throughout the United States.[345]

Transit budgets for large urban areas have a different problem. Most large transit systems have increased suburban-central city commuter services, to help contend with congested highways. However, due to chronic budget tightness since 1982, the spreading of transit has come at the expense of intensive local and crosstown services that sustain walkable neighborhoods in the central cities. "Regionalization" of transit has mustered suburban financial support at the cost of eroding the taxpayer benefits of transit in central city neighborhoods.

In recent years, notably in New York City and Los Angeles, local controversy has erupted over transit's pattern of shifting its resources to suburban commuters. In New York City, the transit authority was sued for allegedly raising downtown subway fares by a higher percentage than commuter train fares. The Los Angeles transit authority was sued over the use of local tax revenues to build suburban-oriented rapid transit at the expense of downtown bus services. These controversies are symptoms of a deeper problem that the approach taken in this book may help to solve.

Transit policy makers are forced into recurring zero-sum (win-lose) decisions by the lack of funding increases. Budgetary stinginess is a direct result of the failure to measure the value of diverse transit benefits. Generally, transit benefits are viewed inaccurately in two ways. First, the value of transit is generally considered to be low, an attitude reinforced by the absence of efforts to value its daily benefits to passengers, constituents, and the general public. The transit industry is beleagured by an assumption that transit is not worth the $20 billion per year that it costs. The transit research agenda, historically reoccupied with costs, has done very little until recently[346] to challenge this underlying attitude.

Second, important distinctions in the value and performance of transit in its diverse functions, if they are considered at all, are blurred. One reason is the generally accepted view that transit trips are of little value to begin with. Another is the inevitable exercise of political influence in the deployment of services and allocation of subsidies. As a result of low implicit valuation and the exercise of political influence, services are cut or expanded in the name of a dubious "feel good" measure, "ridership", with little effort to compare costs against benefits. Ironically, benefit cost analysis of transit budget alternatives, as suggested here, would dramatically change the image of transit's value in all its functions. In so doing, benefit cost consideration would sharpen the policy debate and also attenuate ephemeral political influences.

Every local transit budget should come to the legislative branch as a set of spending alternatives. Each alternative should contain estimates of any change in benefits compared to the previous year. Each scenario should include the relative costs and benefits to passengers, to constituents, and to the general public. Ideally, the legislators should be given the data they need to determine the monetary value of the trade-offs they are asked to make.

More importantly, however, the rigorous collection of information on marginal costs and marginal benefits to transit's diverse beneficiaries, could make a strong case for a larger transit budget. This, of course, would reduce the need for cutting valued city services for equally valued suburban services. Instead of buying "buses", legislators would be buying benefits the value of which they could measure in dollars.

During the 1970s, with Federal initiative, local communities throughout the United States transformed their transit companies into public service agencies. Many of the personnel remained in place, ensuring continued technical and managerial competence to operate safe and efficient transit services. The industry was saved and has persisted in bringing Americans valued benefits. A number of cities are even being shaped by their new rapid transit systems. However, we have been slow to recognize the economic value of transit benefits and, as a consequence, transit has been chronically stifled in the same budgetary process than established transit as a core public function. The findings reported in this book strongly suggest that local measurement of transit's benefits to households, constituencies, and the general public could transform transit service agencies once again into powerful influences on the efficiency and therefore the quality of American neighborhoods and urban areas.

Funds and Budgets

Markets are efficient when consumers are able to compare costs to benefits. So, too, are public sector budgets efficient when taxpayers and legislators are able to compare costs to benefits. Oftentimes, of course, the failure of markets to facilitate such comparisons motivates public sector provision of transit, education, and other public services. Transit and education, however, exemplify public sector activities in which markets continue to play a forceful role through consumer-like behavior of their clients.

Since the tax revolts of the late 1970s rattled all levels of government, elected officials have worked to tailor revenue sources to the benefits of public programs. They have made certain services like sanitation, water, and sewage financially self sufficient by assessing cost-based fees on service recipients. Parent associations have become significant sources of volunteer workers and of funds to supplement school budgets.

The transit industry has diversified its funding too, as summarized in Table 6.1. In 1995 fully 24 percent of transit funds came from dedicated income, sales, and property taxes. The Federal government has led the way in earmarking gas tax revenues for transit investments.

Federal gas tax revenues from the Transit Account of the Highway Trust Fund now provide two-thirds of Federal transit funds. However, general revenues from governments and fares continue to provide the lion's share of transit revenues, together accounting for nearly 57 percent of transit revenues.

Table 6.1 Transit Funding, All Sources, 1995

(Millions)	Operating Funds	Capital Funds	Total	Percent
General Sources				
Dedicated*	$5,177	$388.76	$5,565	24.0
General Revenues	$4,208	$2,408	$6,617	28.5
Gas Taxes	$474	$2,875	$3,349	14.4
Fares	$7,662	–	$7,662	33.0
Total	$17,521	$5,672	$23,193	100

*Dedicated taxes on Income, Sales, Property, and toll revenues.

Source: 1995 National Transit Database

The diversification of transit's financial support in numerous local jurisdictions attests to the intuitive recognition of transit's diverse local policy functions. Many local areas assess regional taxes on themselves to build regional rapid transit systems. San Francisco led the way in regional financial planning when it levied a nine county regional tax to support construction of the Bay Area Rapid Transit system in the 1960s. Similar regional arrangements exist in the Seattle, Houston, Dallas-Ft. Worth, Los Angeles, and Washington, D.C. metropolitan areas. Other areas turned to their respective States to take charge of funding and operation of transit services, as in Rhode Island, Minneapolis, Delaware, Boston, Philadelphia, and to a large extent in New York, New Jersey, and Chicago.

These financial arrangements are best described as "burden-sharing" in the recognition that congestion problems addressed by rapid transit are regional in nature, providing benefits to motorists and to the general economy. Regional taxes for transit in particular are accurately portrayed as beneficial to most people in the region because these taxes reduce the costs of congestion. These costs impinge on the travel time of motorists who may never use transit and because lost time seeps into the costs of products and services in the region, congestion costs affect the whole economy.

However, the regional taxes that have been levied to support transit are only rarely motor fuel taxes, which would focus the burden on motor vehicle use. With the significant exception of New York City, which uses vehicle toll revenues to support transit, most levies are on incomes, sales or real property, tax objects that bear only the remotest relation to use of the highway system. Moreover, sales taxes tend to be regressive taxes, shifting a burden that is not only inversely proportional to ability to pay, but also inversely related to use of the crowded highways.

As suggested earlier, a shift to more systematic measurement of transit benefits would translate into more support by virtue of generating support that is focused on transit's diverse functions and their respective benefits. As much as transit funding should be tied to its benefits, it is even more important that transit funding be linked to the economics of transit services—its costs and revenue potential—in each of its policy functions.

Functions and Markets

In its traditional central city markets, when offpeak and weekend patronage produced sufficient revenues to offset the losses associated with extreme peaking at each end of the workdays, transit was able to cover costs from the farebox. This is still true, if rarely taken advantage of, in many neighborhoods and commercial centers where transit is heavily utilized all day and on the weekends. This traditional transit service profile is one that supports walkable neighborhoods, business districts, and campuses. Such "livable" transit services generally do not need subsidies to cover operations because they enjoy a competitive advantage over autos that can be inconvenient to use and expensive to park in such areas.

Furthermore, transit's physical infrastructure should not be funded from fares. The physical infrastructure of high density rapid, commuter, and light rail systems is best considered the property of taxpayers in general, most efficiently constructed and maintained at the expense of taxpayers in general. Property owners, regional economic enterprises, and the workforce have the most stake in the condition and performance of transit facilities in their immediate vicinity and in the regional as a whole. Chapter 2 suggests that middle income households reap more than $10 billion per year by virtue of the auto ownership savings associated with convenient high quality transit.

Low cost mobility services, with more frequent services to more destinations, would be extremely valuable to the general public. On average, this study suggests, low income passengers and their neighbors receive from $10 to $23 billion per year in benefits from the low cost mobility that transit affords their communities. In this case, frequent, round the clock services to all neighborhoods require per passenger subsidies that are larger than most analysts consider reasonable. But most analyses to date have ignored the value of low cost mobility benefits. They have expressed alarm over a $4.00 subsidy when the benefits, on average, appear to exceed $8.00.

When a public transit demand response vehicle is the substitute for an ambulance, the benefit per trip is much higher. So, too, the benefit to society is correspondingly large when the transit passenger is unemployed and on a job search.

The general economic and general welfare benefits that low cost mobility produces are most reasonably funded from general revenues. These funds are not intended to be associated with benefits from special services, but to represent the fair allocation of unassignable social costs to the general population. The direct benefits of low cost mobility to passengers and its diffuse indirect benefits to the community at large are best understood as general benefits of the public purse. A fuller accounting of these general benefits at the local level could prove effective in increasing transit budgets everywhere in the United States where people are without genuine mobility in an auto-oriented economy.

Finally, as has already been mentioned, the Federal government over a 20-year period has gradually recognized transit as fulfilling the functions of making United States highways more efficient. Accordingly, the Congress early on authorized shifting increasing amounts of Highway Trust Fund revenues to the Transit Account. Moreover, since 1991, transit uses have been eligible expenses in the three major Title 23 highway programs: the Congestion Management and Air Quality program, the Surface Transportation Program, and the National Highway System program.

Local authorities have already transferred billions from highway uses to transit investments, in the recognition that often the most effective highway investment is more transit services to dampen peak hour highway travel demand.

Transit services that are allotted to meet peak period travel demand are by far the most deficit-laden services. These services produce transit's largest deficits and its perennial fiscal problems. State, regional and local transportation planners should look to motor fuel revenues to fund the peak period *operating deficits* of transit systems which measurably reduce demand for peak period highway travel. This principle was recognized in the design of the Federal Congestion Pricing Pilot Program enacted by Congress in 1991, in which increased transit services, and the operating costs they produce, were recognized as essential for success in travel demand management through pricing. As they bring systematic benefit measurement to bear, local authorities will find the local taxpayers willingly would support funding peak transit services from motor fuel taxes.

It is fairly certain that market principles will be brought increasingly to bear on the urban transportation systems in the United States Closely associated with market forces are the costs of transportation and its benefits to individuals and to groups. As traditionally obscure costs and benefits come into focus, it is imperative that public sector budgets make aggressive use of this knowledge. Each transportation budget represents an opportunity to consider relatively small and manageable service reallignements in the light of new information on marginal costs and benefits. At the same time, each budget is an opportunity to reallocate the funding burden among the individuals and groups most able and willing to pay.

Investment and Project Planning

Thus far in this Chapter we have emphasized the importance of linking budgets and benefits. In itself, a "budget" in itself is simply a statement of *inputs*, the dollars, the people, the capital and the other resources to be injected into a program. A budget of course also yields *outputs*. A state's budget for education produces, *inter alia,* higher earning power and more satisfying careers for citizens. A health budget delivers lower infant mortality rates, diminished incidence of infectious diseases, and so on. Planners in these and many other sectors strive to report both the inputs *and* outputs when presenting budgets to decision makers.

Transit authorities, on the other hand, are often a lot more specific about the inputs than outputs. The outputs of transit are affordable mobility, congestion management, and liveable community effects already identified in this book. To be sure, decision makers in the transit world possess a strong qualitative understanding of the outputs of public transportation. They demonstrate a keen awareness of the economic and social value of these outputs relative to the budgetary costs of achieving such value. Proof of this understanding and awareness lies in the consistent year after year funding of local transit budgets through the budgetary process. Even so, a qualitative approach to outputs is a blunt planning instrument, especially when budgets are tight and other sectors are becoming ever more competitive and sophisticated in making their case for scarce funding resources with quantitative attention to outputs.

This book counsels planners in local and regional transit authorities to give quantitative attention to transit outputs in two distinct ways, *ex post* and *ex ante*. The *ex post* consideration of outputs involves estimating and reporting the economic and social value of recent transit budgets. The *ex ante* consideration of outputs involves measuring and reporting the prospective benefits of proposed transit budgets going forward.

Ex Post Measurement

Decision makers, and the taxpayers and passengers they serve, recognize that transit delivers value if each dollar budgeted creates more than a dollar's worth of solid economic and social benefit in the community. Some transit authorities do seek to identify and report, from time to time, the "economic impact" of transit, a term which has become synonymous with the jobs and income created by the transit authority. (We prefer the term "enterprise impacts" – see below). Some others seek to portray the amount of congestion relief provided by the system. We are not aware of any cases however in which a transit authority looks comprehensively, quantitatively and consistently at the economic and social effect it has on the community.

A comprehensive taxonomy of economic and social effects has four elements:

> *Livable Community Effects:* Transit induces high density residential and commercial development with related property value and economic productivity effects;

Congestion Effects: Transit helps reduce congestion-related accidents, delays, extra gas and other auto-related expenses, and environmental emissions;

Affordable Mobility Effects: Transit generates a higher economic standard of living for low income households; and it fosters budgetary savings in non-transportation social services; and

Enterprise Effects: As a regional business enterprise, transit generates employment and income for workers and suppliers.

Local authorities can use this taxonomy to estimate the value of transit's ffect in each category. Methods for estimating enterprise impacts based n employment levels and multiplier effects are well known. Throughout is book we have treated these effects as pecuniary benefits that, *ceteris aribus*, would be produced by any localized public expenditures, icluding tax reductions that could increase jobs by stimulating local onsumer demand. Accordingly, we have steered clear of transit's ecuniary benefits when calculating the total benefits of existing services.

For some purposes the use of enterprise effects are desirable. For xample, if a proposed transit project is to be compared with a highway roject for which entreprise effects have been calculated, the estimation of ransit enterprise effects is desirable. Or, if the multiplier effects of two ypes of investments differ, the enterprise or pecuniary effects of both hould be estimated. In lieu of evidence to the contrary, however, the ntreprise effects of public expenditures should be treated as income ransfers incidental to the provision of public benefits, and not the source of ncome.

Diverse methods for estimating effects in the remaining three categories re presented in this book. These are benefits which flow from transit ervices per se and are proportionate to these services. Taken together, uch estimates can give the community real perspective on the economic nd social performance of transit.

An example, based on actual estimates for 1997 for a Regional Transit uthority operating in the mid-western United States , is given in Table 6.2 table like this one provides decision makers, taxpayers and transit users vith insight into the role transit in a way that words alone cannot. The umbers, combined with a brief and lucid explanation of their proper nterpretation, can sharpen stakeholders' perspective on the contribution of ransit to economic and social objectives and, in particular, the *value* of that ontribution.

It is of course tempting to extract a statement of "total value" from the estimates in Table 6.2. In fact it would be wrong to simply add the values in each category since doing so runs the risk of "double-counting" certain benefits whose value is manifest in more than one way (such as the value of time savings, which is manifest in both congestion relief and livable community benefits). But some adding up *is* valid. At a minimum, congestion effects and affordable mobility effects are additive. Taken together, these two effects combined created $84.6 million of value in the RTA region (see Table 6.2), an amount well in excess of the $50 million in transit operating and capital expenses. Enterprise effects can be added in to the extent that transit employees would, in the absence of transit, have been unemployed or employed in lower paying jobs rather than employed in other, non-transit sectors. Thus enterprise effects cannot be counted into the total value statement when the local economy is operating at full employment. In years of sluggishness in the local economy, however, a part of the enterprise effect, perhaps as much as one-quarter, will certainly be additive.

Table 6.2 Regional Transit Authority Impacts, 1997

	Millions
Expenditures	$50.0
Benefits	
Economic Development	$4.0
Congestion Relief	$23.4
Affordable Mobility	$61.2
*Employment and Income (transfers)**	*$152.5*

*Employment and income ("enterprise") effects from expenditures per se are *economic transfers* and should be included in benefit-cost analysis only when comparing other projects which include such transfers.

The additivity of livable community benefits is also a matter of circumstance. As shown in Chapter 5, livable community benefits are valued in relation to the increment of land value attributable to the presence of transit. Where this value arises because transit users elect to live near a station or a bus stop, it would be wrong to add the livable community effect to the affordable mobility and congestion effects since the latter also reflect the value occasioned by transit users. On the other hand, where the increment of land value attributable to transit is greater than that which can be explained solely by the value of passengers' time, it signifies that land values reflect a class of benefits that are different from transportation benefits per se. It is well known, for example, that transit in some districts creates value for non-transit users in the form of greater urbanity, shorter walking distances to parks, shops, and offices, "insurance" for the rare occasion when a car is unavailable, and so on. Chapter 5 shows how non-transportation livable community benefits can be isolated and the increment added to other categories of benefit.

While absolute precision will never be attainable in extracting a total value statement, an understanding of the "adding up" issues combined with care to explain and avert double-counting will ensure that the presentation of *ex post* transit outputs is legitimate and effective as means of informing decision makers and the general public.

Ex Ante Measurement

Looking to the future, assessing the prospective benefits of next year's transit budgets should take place at two levels, the aggregate level and the project level.

At the aggregate, budget-as-a-whole level, the assessment process is much the same as that outlined above under *ex post* assessment. The categories of outputs, or benefits, are the same and the adding-up issues are identical. Indeed, we would expect transit authorities to present decision makers with an historical account of the benefits, as illustrated in Figure 2, with the projected effects of the forthcoming budget (or budget alternatives) alongside. The estimates would not be expected to change radically from one year to the next unless major capital projects or shifts in revenue support are anticipated or included as scenarios.

At the project level, the output categories and adding-up issues still remain the same, but the costs and benefits to be considered are those strictly associated with a given capital project, a specific shift in service quality (a new schedule, for example) or a specific change in policy with regard to revenue support or fares. It is of course the case that some form of cost-effectiveness analysis is often performed at the project level, at least for capital projects. In this book however we are counseling two significant changes in current practice.

The first change relates to the scope of current practice. Scope limitations under current procedures limit the consideration of benefits to those associated with ridership alone. This is self-evident in the conventional evaluation metric "cost per *trip*" or "cost per additional *trip*" (where an additional trip is one that would not have occurred in the absence of the project). In considering the full range of transit outputs, including the provision affordable mobility and the creation of livable community benefits, the scope of project-level evaluations would be broadened considerably.

The second change pertains to the kind of projects targeted for evaluation. In the past, economic evaluation has been limited to passenger-related capital projects, such as the construction of a new rail line or bus-way. Rarely does one encounter the same kind of attention to non-passenger capital projects, such as a new or expanded maintenance facility. Rarer still is the use of economic evaluation to appraise the prospective value of a service-level project, such as a change (whether up or down) in scheduled headways, the provision or removal of bus shelters, the rehabilitation of stations, and so on. And rarest of all is the application of economic evaluation to prospective changes in fares or levels of revenue support.

The key point to be made here is that all of the project categories identified above can and should be compared with one another in terms of the fundamental outputs that transit is intended to deliver. Economic appraisal at the project level can help decision makers facing a budget constraint decide whether to raise fares or reduce headways; whether to build a light rail system or add substantially to the number of bus routes; whether to paint stations or add more peak service. The list of possible comparisons is of course endless. The crux of our argument is that in bringing a comprehensive view of transit outputs and transit projects to the consideration of specific transit spending options, decision makers and the general public will gain a much deeper appreciation of transit's value than they obtain under today's primarily input-oriented budgeting process.

Notes

344 Jacobs, Jane, *Cities and the Wealth of Nations*, (New York: Random House, 1984).

345 The growth in rural and small urban transit services since the 1980s reinforces this assertion.

346 Federal Transit Administration, *Transit 1996 Report: An Update*, (Washington, D.C.: United States Department of Transportation, 1996).

Bibliography

42nd Street Development Project Inc. (1993), *42nd Street Now! Executive Summary*, New York.

AAA (1995), *Your Driving Costs*, 1995 Edition, Automobile Association of America, Washington D.C.

Al-Mosaind, Musaad A., and Kenneth Deuker and James G. Strathman (1994), 'Light Rail Transit Station and Property Values: A Hedonic Price Approach', *Transportation Research Record No. 1466, Transportation Research Board*, Washington D.C.

Allanson, E.W. (1982), *Car Ownership Forecasting*, Gordon and Breach Science Publishers, London.

Almanza, Susana and Alvarez, Raul (1995), 'The Impacts of Siting Transportation Facilities in Low-Income Communities and Communities of Color', *Transportation: Environmental Justice and Social Equity Conference Proceedings*, Federal Transit Administration, Washington D.C.

American Automobile Manufacturers Assoc. (1995), *Motor Vehicle Facts and Figures*, AAMA, Detroit.

American Public Transit Association (1996), 1996 *Transit Fact Book*, APTA, Washington D.C.

American Public Transit Association (1993), *1993 Transit Fact Book*, APTA, Washington D.C.

American Public Transit Association (1995), *1994-1995 Transit Fact Book*, APTA, Washington D.C.

Anas, Alex (1995), 'Capitalization of Urban Travel Improvements into Residential and Commercial Real Estate: Simulations with a Unified Model of Housing, Travel Mode and Shopping Choices', Federal Transit Administration, Washington D.C.

Armstrong R.J. (1994), 'Impacts of Commuter Rail Service as Reflected in Single-Family Residential Property Values', *73rd Annual Meeting of the Transportation Research Board*, Washington D.C.

Arnott, Richard and Small, Kenneth (1994), 'The Economics of Traffic Congestion – Rush-Hour Driving Strategies that Maximize an Individual Driver's Convenience May Contribute to Overall Congestion', *American Scientist*, vol. 82.

Arrow, Kenneth (1963), *Social Choice and Individual Values*, New York: Wiley.

Attoe, Wayne (ed.) (1988), *Transit, Land-Use and Urban Form*, Center for the

Study of American Architecture, University of Austin Press.
Audirac, Ivonne, Anne H. Shermyen and Marc T. Smith, 'Ideal Urban Form and Visions of the Good Life: Florida's Growth Management Dilemma' (1990), *Journal of the American Planning Association*, Autumn.
Bartic, Timothy J. (1988), 'Measuring the Benefits of Amenity Improvements in Hedonic Price Models', *Land Economics*, vol. 64, no 2, May.
Bellah, Robert N. et al. (1985), *Habits of the Heart-Individualism and Commitment in American Life*, Harper and Row, New York.
Bellah, Robert N., (1992) *The Good Society*, Vintage Books, New York.
Blum, Ulric (1997), 'Benefits and External Benefits of Transport: A Spatial View' in *The Full Costs and Benefits of Transportation*, Springer, New York.
Boyer, Christine, *Dreaming the Rational City: The Myth of American City Planning (1983)*, MIT Press, Cambridge.
Braybrooke, David and Lindblom, Charles E. (1961), *A Strategy of Decision* Free Press, New York.
Brod D. (1996), 'Accounting for Multi-Modal System Performance in Benefit-Cost of Transit Investment', *Transportation Research Board*, National Research Council, Washington D.C.
Buchanan, James M (1954a), 'Social Choice, Democracy, and Free Markets, *Journal of Political Economy*, vol. 62(2).
Buchanan, James M. (1963), *Traffic in Towns: A Study of the Long-Term Problems of Traffic in Urban Areas*, London.
Buchanan, James M. (1965), 'An Economic Theory of Clubs', *Economica*, February.
Bureau of the Census (1993), *Statistical Abstract of the United States 1993, 113th Edition*, United States Department of Commerce, Washington D.C.
Button, Kenneth J. (1993), *Transport Economics – 2nd Edition*, Edward Elgar Publishing Ltd., Vermont.
Cambridge Systematics (1995), 'A Review of Methodologies for Assessing the Land Use and Economic Impacts of Transit on Urban Areas.', *FTA Discussion Paper*, FTA, Washington D.C.
Canadian Ministry of State for Urban Affairs (1976), *Creating a Livable Inner-City Community: Vancouver Experience*, Agency Press.
Carr, Melanie et al.(1993), *Cross-sector Benefits of Accessible Public Transport*, Environment Resource Centre, Crowthorne, Berkshire, U.K.
Cervero, Robert (1984), 'Light Rail Transit and Urban Development', *APA Journal*, Spring.
Charles River Associates Inc. (1977), *Public Transportation Fare Policy*, United States Department of Transportation, Washington D.C.
Charles River Associates Inc. (1986), *Allocation of Federal Transit Operating Subsidies to Riders by Income Group*, Federal Transit Administration, Office of Policy Development, Washington D.C.
Chen, Don (1995), 'Social Equity, Transportation, Environment, Land Use and Economic Development: The Livable Community', *Environmental Justice and*

Social Equity Conference Proceedings, Federal Transit Administration Washington D.C.

'Congress Clears $375 Million Mass Transportation Bill', *Congressional Quarterly*, July 3, 1964.

Cudahy, Brian J. (1982), *Cash, Tokens, and Transfers: A History of Urban Mass Transit in North America*, New York: Fordham.

Delucci, Mark A. (1996), *The Annualized Social Cost of Motor-Vehicle Use in The United States. 1990-1991: Summary of Theory, Data, Methods and Results*, Institute of Transportation Studies, Davis, CA.

Directions: The Final Report of the Royal Commission on National Passenger Transportation, (1992), Minister of Supply and Services Pub., Ottawa, Canada, vol. 2.

Dittmar, Hank and Chen, Don (1995), 'Equity in Transportation Investments', *Environmental Justice and Social Equity Conference Proceedings*, Federal Transit Administration, Washington D.C.

Downs, A. (1962), 'The Law of Peak-Hour Expressway Congestion', *Traffic Quarterly*, vol. 16.

Downs, Anthony (1992), *Stuck in Traffic- Coping with Peak-Hour Traffic Congestion*, The Brookings Institute, Washington D.C.

Edgeworth F.Y. (1925), *Papers Relating to Political Economy*, London.

Federal Highway Administration (1992), *New Perspectives in Commuting*, United States Department of Transportation, Washington D.C.

Federal Transit Administration (1994), *Livable Communities Initiative: Program Description*, United States Department of Transportation, Washington D.C.

Federal Transit Administration (1994), *Unsticking Traffic: When Transit Works and Why*, United States Department of Transportation, Washington D.C.

Federal Transit Administration (1996), *Transit 1996 Report- An Update*, United States Department of Transportation, Washington D.C.

Fishman, Robert, *Bourgeois Utopias: The Rise and Fall of Suburbia* (1987), Basic Books, New York.

Gantzlaff, Dean H. and Smith M. (1993), 'The Impact of the Miami Metrorail on the Value of Residences Near Station Locations', *Land Economics*, vol. 69, no. 1, February.

Glaister S. and Lewis D. (1978), 'An Integrated Fares Policy for Transport in London', *Journal of Public Economics*, vol. 9.

Glaister, Stephen, (1981), *Fundamentals of Transport Economics*, Basil Blackwell, Oxford.

Goodwin, P.B. (1969), "Car and Bus Journeys to and from Central London in Peak-Hours', *Traffic Engineering and Control*, Vol. 11.

Gordon, Stephen P. And John B. Peers, 'Designing a Community for Transportation Demand Management: The Laguna West Pedestrian Pocket' (1991), (in) *Transportation Research Record No. 1321*, Transportation Research Board 1321, Washington D.C.

Green R.D. and James O.M. (1993), *Rail Transit Station Area Development: Small Area Modeling in Washington D.C.*, M.E. Sharp, New York.

Hall, Peter, *Cities of Tomorrow: An Intellectual History of Urban Planning and Design in the 20th Century* (1988), Basil Blackwell, London.

Hensher, David et al. (1988), 'The Role of Stated Preference Methods in Studies of Travel Choice', *Journal of Transport Economics and Policy*, January.

Hickling Lewis Brod Economics (1995), 'Basic Mobility Needs and Related Transit Activities', *FTA Working Paper 1.1*, Ref. 5273-300.

Hillman, Mayer et al. (1973), 'Personal Mobility and Transport Policy', *PEP*, vol. 34, no. 34, June.

Hodges, Luther H. (1962), 'Urban Transportation-Joint Report to the President by the Secretary of Commerce and the Housing and Home Finance Administrator' *Hearings on the Urban Mass Transportation Act of 1962* (H.R. 11158), House Committee on Banking and Currency, Washington D.C.

Holtzclaw, John (1991), 'Explaining Urban Density and Transit Impacts on Auto Use', *California Energy Commission Docket No. 89-CR-90.*

Holtzclaw, John (1994), 'Using Residential Patterns and Transit to Decrease Auto Dependence and Costs', *Natural Resources Defense Council*, June.

Hotelling, H. (1932), 'Edgeworth's Taxation Paradox and the Nature of Supply and Demand Functions', *Journal of Political Economy*, vol. 40, no. 5.

Hughes, Mark A. (1989), *Fighting Poverty in Cities- Transportation Programs as Bridges to Opportunity*, National League of Cities, Washington D.C.

Hunt, J.D. et al. (1994), 'A Stated Preference Investigation of Influences on the Attractiveness of Residential Locations', Transportation Research Board Conference.

Jacobs, Jane (1960), *The Economy of Cities*, Random House, New York.

Jacobs, Jane (1961), *The Death and Life of Great American Cities*, Vintage Books, New York.

Jeanneret, Charles-Edouard (aka Le Corbusier) (1964), *When Cathedrals Were White*, New York: McGraw-Hill, 1964.

Johnson, Elmer (1994), 'Collision Course: Can Cities Avoid A Transportation Pileup?", *American City and Country*, vol. 109, no. 3, March.

Jones, David (1985), *Urban Transit Policy: An Economic and Political History*, Engelwood Cliffs, New Jersey.

Kain, John, 'Housing Segregation, Negro Employment, and Metropolitan Decentralization' (1968), *Quarterly Journal of Economics*, vol. 82.

Kain, John F. (1992), 'The Spatial Mismatch Hypothesis: Three Decades Later', *Housing Policy Debate*, vol. 3, Issue 2. Fannie Mae, Office of Housing Policy Research.

Kain, John F. (1992), 'Impacts of Congestion Pricing on Transit and Carpool Demand and Supply', in *Curbing Gridlock – Peak-Period Fees to Relieve Traffic Congestion*, vol. 2 , National Research Council, Washington D.C.

Katzmann, Robert (1986), *Institutional Disability: The Saga of Transportation Policy for the Disabled*, The Brookings Institution, Washington D.C.

Kitamura, Ryuichi, 'Life-Style and Travel Demand', 1988, (in) *Special Report 220: A Look Ahead,* Transportation Research Board, National Research Council.

Knight, Richard V., 'Changes in the Economic Base of Urban Areas: Implications for Urban Transportation (1983), (in) *Future Directions of Urban Public Transportation,* Transportation Research Board Special Report 199, Washington D.C.

Kunstler, James Howard, *The Geography of Nowhere: The Rise and Decline of America's Man-Made Landscape* (1993), Simon and Schuster, New York.

Laffont J.J (1989), *Fundamentals of Public Economics,* MIT Press, Cambridge.

Lakshmanan T.R., Nijkamp Peter, Verhoef Erik (1997), 'Full Benefits and Costs of Transportation: Review and Prospects' in *The Full Costa and Benefits of Transportation*, Springer, New York.

Landis, John and Cervero, Robert (1995), 'BART at 20: Property Value and Rent Impacts', *74th Annual Transportation Research Board,* January.

Landis, John et al. (1995), 'Rail Transit Investment, Real Estate Values, and Land Use Change: A Comparative Analysis of Five California Rail Transit Systems', *Institute of Urban and Regional Development,* University of California at Berkeley, July.

Layard, Richard and Stephen Glaister (1994), *Cost-Benefit Analysis,* Cambridge University Press, Cambridge.

Lerman, Steve R. et al. (1978), *The Effect of the Washington Metro on Urban Property Values,* Urban Mass Transportation Administration, July.

Lewis, David (1975), 'Public Transport Fares and the Public Interest', *Town Planning Review,* vol. 46, no 3 July.

Lewis, David (1977), 'Estimating the Influence of Public Policy on Road Traffic Levels in Greater London' *Journal of Transport Economics and Policy,* May.

Lewis, David (1985), *The Economics of Serving The Travel Needs of the Handicapped Persons in the United States,* London School of Economics, London.

Lewis, David (1978), 'Public Policy and Road Traffic Levels: A Comment', *Journal of Transport Economics and Policy*, vol. XII, no 1, January.

Lewis, David (1982), 'Transportation for Handicapped Persons: From Policy to Administration', *Specialized Transportation Planning and Practice*, vol. 1, no 1.

Lewis, David (1982), *The Interstate Highway System: Issues and Options,* Congressional Budget Office, Washington D.C.

Lewis, David (1987), 'Making Paratransit Service Level Decisions When Budgets are Constrained', *Specialized Transportation Planning and Practice*, vol. 6, no 4.

Lewis, David, Daniel Hara and Joseph Revis (1988), 'The Role of Public Infrastructure in the 21st Century', (in) *Special Report 220: A Look Ahead, Transportation Research Board,* National Research Council.

Lewis, David and Ling Suen (1989), 'Towards a Doctrine of Mobility as a Human Right', (in) *Proceedings of a Conference held at Stockholm, Sweden of the*

Swedish Board of Transport in Cooperation with the Department of Traffic Planning and Engineering, Lund Institute of Technology, May 21-24.

Lewis, David and Mark Haney (1989), 'Efficiency, Innovation, and Productivity in Public Infrastructure', (in) *Proceedings; Second Canadian Conference on Urban Infrastructure,* February 13 and 14.

Lewis, David (1995), 'The Future of Forecasting: Risk Analysis as a Philosophy of Transportation Planning', *TR News,* no 177, March-April.

Lewis, David and Michael O'Connor (1997), 'Economic Value of Affordable Mobility', (in) *Proceedings of the Transportation Research Board*, paper no 1, 97-1093, January.

Lewis-Workman, Steven and Daniel Brod (1997), 'Measuring the Neighborhood Benefits of Rail Transit Accessibility', (in) *Transportation Research Board,* paper no 97-1371 January.

Lewis, David (1998), 'Impact of Reliability on Paratransit Demand and Operating Costs', *Transportation Planning and Technology*, vol. 21, no 4.

Lowe, Marcia, Alternatives to the Automobile: 'Transport for Livable Cities' (1990), World Watch Institute Paper 98, Washington D.C.

Lindblom, Charles E. (1977), *Politics and Markets*, Basic Books, New York.

Maddala, G. S., (1983), *Econometrics,* McGraw-Hill Inc.

Mackenzie James J. et al. (1992), *The Going Rate: What It Really Costs to Drive*, World Resources Institute, Washington D.C.

Mantell, Michael, Stephen T. Harper and Luther Propsi, *Creating Successful Communities: A Guidebook to Growth Management Strategies* (1990), The Conservation Society, Island Press, Washington D.C.

Meyer, John and Gomez-Ibanez, Jose (1981), *Autos, Transit and Cities*, Harvard, Cambridge.

Meyer J.R., Kain J.F., and Wohl, M. (1965), *The Urban Transportation Problem,* Harvard University Press, Cambridge.

Meyerhoff, Alex et al. (1993), 'Running on Empty: Travel Patterns of Extremely Poor People in Los Angeles', *Transportation Research Record 1395,* Transportation Research Board, Washington D.C.

MillerPeter and Moffet, John (1993), *The Price of Mobility,* Natural resources Defense Council, San Francisco.

Moffet, John and Peter Miller, *The Price of Mobility,* Natural Resources Defense Council (pub.), San Francisco.

Mogridge, Martin J.H. (1990), *Travel in Towns*, MacMillan, London.

Moore, Terry and Thorsnes, Paul (1994), 'The Transportation/Land Use Connection: A Framework for Practical Policy', *Report Number 448/449*, American Planning Association, Washington D.C.

Mumford, Lewis (1938), *The Culture of Cities*, Harcourt Brace, New York.

Mumford, Lewis (1961), *The City in History: Its Origins, its Transformations and its Prospects,* Harcourt, Brace and World, New York.

Musgrave, R.A. (1959), *The Theory of Public Finance*, McGraw-Hill, New York.

Oram, Richard L. (1979), 'Peak Period Supplements: The Contemporary Economics of Urban Bus Transport in the U.K. and the USA", in *Progress in Planning*, vol. 12, Part 2, Pergamon Press, Oxford.

Oram, Richard L. (1988), *Deep Discount Fares: Building Transit Productivity with Innovative Pricing*, United States Department of Transportation, Washington D.C.

Pickrell, Donald H. (1992), 'A Desire Named Streetcar', *Journal of the American Planning Association*, vol. 58 no 2, Spring.

Pickrell, Donald H. (1983), *The Causes of Rising Transit Operating Deficits,* John F. Kennedy School, Harvard University, Cambridge, Massachusetts.

Pigou, A.C. (1920), *The Economics of Welfare*, MacMillan, London.

Pucher, John (1982), 'Discrimination in Mass Transit', *Journal of the American Planning Association*, Summer.

Pucher, John and Williams, Fred (1992), 'Socioeconomic Characteristics of Urban Travelers: Evidence from 1990-1991 NPTS', *Transportation Quarterly*, vol. 46, no. 4, October.

Pushkarev, Boris and Jeffrey Zupan (1980), *Urban Rail in America: An Exploration of Criteria for Fixed-Guideway Transit,* U.S. Department of Transportation.

Rappoport, Amos (1975), 'Toward a Redefinition of Density', *Environment and Behavior*', vol. 7, no. 2, June, 1975), pp. 133-58.

Real Estate Research Corporation (1974), *The Costs of Sprawl: Vol. 1,* United States Government Printing Office, Washington D.C.

Roseland, Mark (1992), *Toward Sustainable Communities – A Resource Book for Municipal and Local Governments,* National Round Table Series on Sustainable Development, Ottawa, ON.

Rybczynski, Witold, (1995), *City Life: Urban Expectation in a New World,* Scribner, New York.

Schrank D.L., Turner S.M., and Lomax T.J. (1993), *Estimates of Urban Roadway Congestion – 1990,* United States Department of Transportation, Washington D.C.

Sen, Amartya (1993), 'Welfare, Resources, and Capabilities: A Review of Inequality Reexamined', *Journal of Economic Literature,* December.

Sen, Amartya (1995), 'Rationality and Social Choice", *The American Economic Review",* March.

Sharp, C.H. (1967), 'The Choice Between Cars and Buses on Urban Roads', *Journal of Transport Economics and Policy,* vol. 1.

Simon, Herbert A. (1976), *Administrative Behavior, Third Edition*, Free Press, New York.

Slater, Rodney E. (1997), *Report on Funding Levels and Allocations of Funds for Transit Major Capitol Investments*, United States Department of Transportation, Washington D.C.

Smeed R.J. and Wardrop J.G. (1964), 'An Exploratory Comparison of the Advantages of Cars and Buses for Travel in Urban Areas', *Journal of the Institute of Transport*, vol. 30, no. 9.

Smerk, George M. (1976), 'Productivity and Mass Transit Management', in *Urban Transportation Efficiency*, ASCE, New York.

Stiglitz, Joseph E. (1986), *Economies of the Public Sector*, W.W. Norton and Company, New York.

Suchorzewski, W. (1973), 'Principles and Applicability of the Integrated Transportation System', Proceedings of the UN-ECE Seminar on the Role of Transportation in Urban Planning, Development and Environment, Muenchen, Germany.

Sugden, Robert, *The Political Economy of Public Choice*, Martin Robertson Pub., Oxford.

United States Bureau of Statistics (1996), *National Transportation Statistics*, United States Department of Transportation, Washington D.C.

United States Department of Transportation (1988), *The Status of the Local Mass Transportation: Performance and Condition*, Report to Congress, June, Washington D.C.

United States Department of Transportation (1988), *Transit Profiles: Agencies in Urbanized Areas Exceeding 200,000 Population*, United States Department of Transportation, Washington D.C.

United States Department of Transportation. *1990 Nationwide Personal Transportation Survey*.

United States Department of Transportation. *1995 Nationwide Personal Transportation Survey*.

United States House of Representatives (1963), *Committee on Banking and Currency on HR 3881, Hearings*, United States Government Printing Office, Washington D.C.

United States House of Representatives (1963), *Transit Development Program for the National Capital Region, Report no. 1005*, United States Government Printing Office, Washington D.C.

United States House of Representatives (1963), *Transit Program for the National Capital Region, Hearings*, United States Government Printing Office, Washington D.C.

United States House of Representatives (1962), *Urban Mass Transportation Act of 1962, Hearings*, United States Government Printing Office, Washington D.C.

United States House of Representatives (1963), *Urban Mass Transportation Act of 1963, Hearings*, United States Government Printing Office, Washington D.C.

United States House of Representatives (1966), *Committee on Appropriations on Transportation Related Agencies, Appropriations for 1964, Hearings*, United States Government Printing Office, Washington D.C.

United States House of Representatives (1996), *Committee on Appropriations on Transportation Related Agencies, Appropriations for 1997, Hearings*, United States Government Printing Office, Washington D.C.

The Urban Institute and Cambridge Systematics (1991) *The Economic Impacts of SEPTA on the Regional and State Economy*, United States Department of Transportation, Washington D.C.

Urban Mass Transportation Administration, *Demographic Change and Recent Work trip Travel Trends*, 1985.

Urban Mass Transportation Administration, *Transit Performance and Needs Report*, 1987.

Voith, Richard (1991), 'Transportation, Sorting and House Values', *AREUEA Journal*, vol. 19, no. 2.

Webster F.V. and Oldfield R.H. (1972), *A Theoretical Study of Bus and Car Travel in Central London*, Transport and Road Research Laboratory, Crowthorne, Berks, U.K.

Whyte, William H. (1988) *Rediscovering the Center,* Anchor Books, New York.

Index

For Product Safety Concerns and Information please contact our EU representative GPSR@taylorandfrancis.com Taylor & Francis Verlag GmbH, Kaufingerstraße 24, 80331 München, Germany

T - #0096 - 160425 - C0 - 215/145/16 - PB - 9780367000011 - Gloss Lamination